LAST ANIMALS
AT THE ZOO

A Shearwater Book

LAST ANIMALS
AT THE
ZOO

HOW MASS EXTINCTION
CAN BE STOPPED

COLIN TUDGE

ISLAND PRESS
Washington, D.C. · Covelo, California

ABOUT ISLAND PRESS

Island Press, a nonprofit organization, publishes, markets, and distributes the most advanced thinking on the conservation of our natural resources—books about soil, land, water, forests, wildlife, and hazardous and toxic wastes. These books are practical tools used by public officials, business and industry leaders, natural resource managers, and concerned citizens working to solve both local and global resource problems.

Founded in 1978, Island Press reorganized in 1984 to meet the increasing demand for substantive books on all resource-related issues. Island Press publishes and distributes under its own imprint and offers these services to other nonprofit organizations.

Support for Island Press is provided by Apple Computer, Inc., Geraldine R. Dodge Foundation, The Energy Foundation, The Charles Engelhard Foundation, The Ford Foundation, Glen Eagles Foundation, The George Gund Foundation, William and Flora Hewlett Foundation, The Joyce Foundation, The John D. and Catherine T. MacArthur Foundation, The Andrew W. Mellon Foundation, The Joyce Mertz-Gilmore Foundation, The New-Land Foundation, The J. N. Pew, Jr. Charitable Trust, Alida Rockefeller, The Rockefeller Brothers Fund, The Rockefeller Foundation, The Florence and John Schumann Foundation, The Tides Foundation, and individual donors.

First Island Press Edition, 1992

© Colin Tudge 1991

Library of Congress Cataloging-in-Publication Data
Tudge, Colin.
    Last animals at the zoo : how mass extinction can be stopped/
Colin Tudge.
  p. cm.
    Includes bibliographical references and index.
    ISBN 1-55963-158-9 (alk. paper)
    1. Wildlife conservation. 2. Endangered species—Breeding.
3. Zoo animals—Breeding. 4. Captive wild animals—Breeding. I. Title
QL82.Y83 1992
639.9—dc20   91-39773
                                  CIP

Printed on recycled, acid-free paper

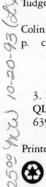

Manufactured in the United States of America
10 9 8 7 6 5 4 3 2 1

CONTENTS

Acknowledgements

So many scientists, directors, curators and keepers have given me their time this past two decades that I hesitate to mention any by name, for fear of omitting some I should have mentioned. I am very grateful to them all. In particular, however, I am aware of my debt to my friends and associates at Chester Zoo, at the Zoological Society of London, and within the Federation of British Zoos, with whom I have had the pleasure of working over the past year or so. I am grateful, too, to Dr Sophie Botros, who is not a scientist, but is an excellent moral philosopher from London University, who helped me to sort out my ideas in Chapter 1.

I am especially aware of my debt also to my fellow biologists who are also writers and editors, and have helped me in many ways over many years: Dr Donald Gould, the most mellifluous turner of phrases, who gave me my first writing job; Dr Michael O'Donnell, who paid me a perfectly good salary while I wrote my very first book; Graham Chedd and the late Ian Low, who started my career at *New Scientist* in the early 1970s; and Dr Bernard Dixon, Dr Roger Lewin, Dr Jeremy Cherfas, and Miranda Robertson, who have been unfailingly helpful and supportive, in some cases (I am chastened to find) over more than 20 years.

Finally, I would like to thank Ann Scott, who read the present text so very professionally, and made my life a lot easier; and Neil Belton of Hutchinson Radius, the most patient and supportive of editors.

INTRODUCTION

The thesis of this book is simple: that zoos are now an essential part of modern conservation strategy; and that of the several tasks that fall to them, by far the most important is the breeding of endangered animals. In short, this book is about captive breeding – or, as I like to call it, 'conservation breeding.'

Surely, though, it is obvious that zoos can save animals by breeding them? Surely this is not worth writing a book about?

It certainly has not been obvious for very long, and it is not obvious to every senior zoo person even now. Until recently, curators tended to list 'conservation' just as one item in a list of three of the principal concerns of zoos, the others being 'education' and 'entertainment'. Only in recent decades has it been realised – only in recent decades has it become true – that for an increasing number of species, the populations inside zoos are larger than those of the wild. In addition, the zoo populations are often far safer than those of the wilderness, and in many cases are growing, while the precarious populations of the wild continue to dwindle. In all but a very few cases, too, the populations in legitimate zoos are growing because the animals are breeding; not because the zoos are 'robbing' the wild. When serious conservationists do take animals from the wild these days, it is to rescue them when extinction is otherwise inevitable. The California condor, red wolf, and Arabian oryx are perhaps the most famous examples so far. Whenever animals are clearly 'doomed' – reduced to a few bewildered individuals – a round-up should at least be contemplated.

Of course the ideal is to save what is left of the wilderness; to protect the wild places where animals live. *Of course*. But in the short term that is not always an option. Always there are countries at war; always there are seemingly inexorable plans to change the landscape, with dams and harbours and cities; still the human population grows as no population of large animals has ever grown in the past. This expansion seems bound to continue for at least several decades, and the human population will remain at an enormous level for several centuries to come. Then again, most of the world has already been compromised. The remaining places that are recognisably pristine are far smaller than the continents of which they are a part, and the ecological forces that obtain in small patches of land are quite different from those of continents. Even in times of peace and plenty, then, it is no longer easy to conserve the 'wild'. It is not enough to put up a fence, and a 'Keep Out' sign, and call it a 'reserve'. Wilderness has at least to be protected and, generally, managed, with varying degrees of intensity. In turbulent times (and these are certainly turbulent times, biologically and politically) the necessary steps cannot be taken. If they are not, then the animals become extinct. If the animals become extinct now, then they will be extinct forever.

Conservation by breeding in zoos is for some animals more feasible than the protection of habitats. For an expanding register of large vertebrates in particular – various rhinos, various subspecies of tiger, a growing list of primates, several cranes, many parrots, various reptiles – it has become the *only* option with a reasonable chance of success in the short and medium term. If we do not take appropriate steps in the short term, then we can forget the long term. What is gone, is gone.

But although breeding in captivity for conservation is feasible – maintaining viable populations of animals for generation after generation – it is far from simple. True, the more obliging species have been producing the odd zoo baby for many a year. That was how zoos kept the visitors flocking. But the occasional baby from the most fecund creatures is not what is called for. Reproduction must be reliable, in the most recalcitrant species as well as in the complaisant ones; and when it is reliable, it must be 'managed' to avoid inbreeding and other genetic disasters. The theory that allows such management is new – it was first formulated in a recognisably modern form only in the 1970s. Both the theory and the necessary techniques (including those of molecular biology and of reproductive physiology) are still evolving.

Still, though, we might argue – and many a conservationist does argue – that if animals can exist only in zoos, then they are better off dead. If the animals' opinions were asked, they might not agree. At least, they might agree if the zoo was of the hideous, old-fashioned kind, a row of boxes with bars. But modern zoos are not like that; or, at least, the best parts of the best modern zoos are not like that, and those that are still behind the times are in many cases restructuring as rapidly as possible. Zoos, as I will discuss towards the end of this book, are beginning more and more to simulate the wild, and increasingly are allowing animals to live as they would do in their proper setting.

The zoo, though, however agreeable it may become, is not the end of the road. The aim of serious conservation breeders is to return animals to their native habitats, as soon as those habitats can be made safe for them again. You may feel that this can never happen. But even now, here and there, circumstances are improving. Britain has already decided that it does not need quite as much land as once it did for agriculture, for modern farming is extremely productive. The UK is 'setting aside' enormous areas, some of which at least could be given to wildlife. Indeed there is room in Britain, even now, to reintroduce the big animals that lived here until medieval and even until modern times – the wolf, the boar, and possibly even the bear.

What is needed in addition to land, of course, is a change of mind; but people do change their minds. The people of the Carolinas in the United States, who once put a bounty on every wolf, are now making space for the red wolves, which have been saved from extinction by breeding in captivity (in Washington state). The Arabian oryx once roamed an area the size of India, yet was hunted to extinction by the early 1970s. A few were rescued, and placed in Phoenix Zoo, Arizona. In the late 1970s the Sultan of Oman said that he would like the oryx to return, and that his people, the Harasis, would look after them; and now there are quasi-wild herds in Oman, Jordan, Saudi Arabia, and Israel. Worldwide, indeed, about 100 'reintroduction' schemes are already in train. In a few centuries' time – and in conservation we must think in centuries – the human population might again begin to diminish, so modern demographic theory suggests. When it does, reintroduction can begin *en masse*. But it cannot begin at all, unless we save the animals now.

But if the thesis is simple, why is the book so long?

I did not intend it to be quite so long. Indeed, when I began writing it in 1976 (I remember the date, because I wangled a trip to Hanover so as to look at the zoo, and justified this because, I said, I was 'writing a book') I did not realise that it would not be simply about the mechanics of husbandry and of display, but about the theory and practice of modern conservation. But then, in 1976, very few people did realise that zoos would come to play such a part as this. The book is long because there is a lot to say. The theory is complicated (extremely interesting, I hope; but complicated). The practicalities I find endlessly fascinating, for every animal is different. Each one poses a new problem. Each is precious, and time is short.

The structure of the book, at least, is simple. The theory and practice of conservation breeding – husbandry, genetics, and what is actually being done – occupy the middle: chapters 3, 4, and 5. The last three chapters, 6, 7, and 8, look towards the future: modern and future reproductive technologies; the enrichment of life in zoos in order to improve the welfare of the animals and to prepare them for the wild; and the future of zoos through the next few centuries.

We begin, though, by addressing more fundamental issues. In chapter 2, we discuss the present state of animals, and ask what zoos (and other kinds of reserve) can really do to help them. In chapter 1 we ask the most basic question of all. Why bother?

ONE

Why Conserve Animals?

There may be 30 million different creatures on this Earth; 30 million different *species*, that is, each kind capable of breeding successfully only with others of its own kind, and not with those who are not. Most of those 30 million live in the tropical rainforests. Most of them are animals. Most of those animals are insects. And most of those insects are beetles – for God, as J. B. S. Haldane allegedly remarked, had 'an inordinate fondness for beetles'.

What use are they, most of them? The figure of 30 million is an estimate, for only 1 million or so have yet been counted. Most of those that have been described are beetles, too (which is why biologists believe that most of those that have not been described are beetles), and only a handful of people in all the world can tell most of them apart. Many, we know, are disappearing by the hour, because the forests in which they live are being cleared away, and they do not live anywhere else. What care I, what care you, what difference does it make to anybody? What difference does it make, come to that, even if more spectacular creatures go by the board? When did you last see a Hyacinthine macaw, that you should regret its passing? What have Sumatran tigers ever done for us?

Zoos justify their existence these days – or at least the serious ones, which are the subject of this book, do – by their contribution to animal conservation. But why should they bother? Why should any of us give a damn?

It may well be true that Sumatran tigers and Hyacinthine macaws seem to contribute very little to our daily lives. There is, though, a strong group of arguments which in a general way we might call 'utilitarian', which say that wild animals and plants can be good for us, and that this is a good reason to hang on to them. What are these arguments?

'WHAT'S GOOD FOR ANIMALS IS GOOD FOR US'

The International Union for the Conservation of Nature and Natural Resources is affiliated to the United Nations and is in effect the 'offical' organisation and voice of conservation worldwide. IUCN has declared its belief that conservation policies should seek to reconcile the needs of wildlife with those of people.

This approach is far from foolish. There is first of all the inescapable reality: that the human population passed the five-billion mark in the 1980s, and that it will probably peak some time in the mid-21st century at between eight and 12 billion, and remain at this prodigious height (if the world can sustain it) for centuries to come. Michael Soulé, a leading American ecologist who is president of the Society for Conservation Biology, envisages a future world that is like South-east Asia today, 'where every square inch is put to use'. Such a mass of people will not be denied. Unless they see some benefit from the wildlife in their midst, they will destroy it, and no amount of legislation will prevent them. In parts of Africa people are shot on sight for poaching; throughout history, indeed, poaching has commonly been treated as a capital crime. But each person can die only once; and if people are threatened with starvation, then they risk the retribution.

Besides, most of the world's leading conservationists (including Dr Soulé) are humanitarians. Some people have said in their time that they 'prefer animals to people'; but I personally know no one who wants to see people disenfranchised, even in the worthiest of conservation causes. Conservation cannot work but also should not work, unless it also takes account of the needs of people. It is indeed incumbent on us to seek compatibility.

Fortunately, there are many ways in which the needs of people and the needs of animals can be reconciled. The utilitarian arguments make an impressive list. For a start, one of the biggest industries these days is tourism; human beings have become the most mobile of creatures, to add to their list of 'firsts'. For rich countries and poor alike, but particularly for poor countries, tourism has become a prime source of income. Kenya derives a third of its income from tourists. Richard Leakey, director of the Kenya Wildlife Services, has no doubt that the principal reason for going to Kenya is to look at the animals. The country has wonderful beaches, to be sure, but so do a lot of other places. Only the countries of Africa have retained the Pleistocene megafauna – the big mammals – in such abundance. Allow them to die and then, he says, the Kenyan economy would be shot to pieces. Rwanda, too, an even poorer country to the north of Kenya, makes money from its mountain gorillas. Few outsiders would even have heard of the country, and even fewer would go there, were it not for them. Rich countries also 'sell' their wildlife. The people of Queensland gain from the tourists who come to admire the Great Barrier Reef; and indeed it has been suggested that the overall management of the Reef (which includes provision for science and fishing, as well as tourists) is a model for wilderness everywhere.

People want more than photographs and memories. They want trophies too, even if only a seed-case or a shell; and they are prepared to pay for them. This raises many issues, both ethical and practical. The bottom line, though, is that every successful population of creatures produces more offspring than its environment can contain. Some individuals *should* be culled in the interests of conservation; and if they are not, they will die anyway. In general, then, it makes sense to sell 'products' from the animals that the habitat cannot support. To this end, IUCN suggests that wild habitats can be modified somewhat, to increase the output. Thus it supports schemes in Papua New Guinea to plant extra food trees in the forest, to increase the population of large, colourful butterflies of the kind favoured by tourists. Without such a trade, the IUCN argues, the local people have no option but to cut the forest down, and grow crops. The raising of butterflies gives them a reason to protect the forest. Some African countries, such as Zimbabwe, charge very rich people appropriately enormous sums to shoot their surplus elephants, and other 'big game'; animals that would have to be culled in any case.

It has often been pointed out, too, that animals native to a difficult territory may fare much better than livestock imported from elsewhere. In particular, antelope and zebra thrive in the African savannah, where domestic cattle, even of the tropical zebu kind, often find themselves in desperate straits. There has been many a scheme, then, to cull small antelope for meat, and even to milk the eland, which is the largest antelope of all.

A more general argument along similar lines – one much favoured by Dr Norman Myers, which he applies in particular to the rainforest[1] – is that wild creatures may have all kinds of properties, and contain all kinds of genes, which may not be useful to us now, but which could be useful in the future. We do not *know* that they will be useful; but that is the point. All we can say is that many wild creatures have already proved useful, and as we have so far identified only a small percentage of what the world contains – and examined precious few of those in detail – we can be sure that there is much more to be found and exploited. Discoveries so far include most of the world's most popular drugs, from aspirin to agents in the Madagascar periwinkle that are effective against leukaemia; and many a gene contained within the wild relatives of food and other crops, which help confer resistance against disease, or drought, or salinity. To be sure, most such examples come from plants, which are the world's most accomplished pharmacologists. But animals are sources of such agents too; including corals from Australia, which are exposed to the sun all day, and produce protective materials which manufacturers of suntan lotions are now extracting and synthesising.

Finally, it is widely argued that biodiversity helps in a general way to conserve the stability of the world's ecology; and hence to maintain a generally benign environment for all of humanity as well as for the rest of life. A typical version of this argument runs as follows. Tropical forests absorb huge quantities of carbon dioxide, and if the trees are removed the CO_2 content of the atmosphere increases, which enhances the greenhouse effect; the stability of the tropical forests depends upon their enormous biological diversity; so if this diversity is reduced, the existence of the forests is placed in jeopardy, and we could all suffer from consequent global warming.

These arguments seem powerful. Are they?

IN PRAISE OF EXPLOITATION

Of course, the attempts to exploit animals in order to guarantee their survival raise many problems. The obvious ethical problems do not bother me too much. It is difficult in this world to do anything that is unequivocally good, and if we have to choose between exploitation and obliteration, then the former emerges as the lesser of the two evils. Besides, the animals that are sold as trophies are, in general, those that would have to be culled in any case; and it is ethically no worse to kill and eat an antelope than to kill and eat a domestic sheep. At least, the sheep would not appreciate the difference.

The practical problems are far greater. For example, at the time of writing (early 1991) Richard Leakey is trying to raise an enormous amount of money from the world at large (about US$200 million) to upgrade the facilities that Kenya provides for tourists. At present the visitors threaten to destroy what they have come to see. Cheetahs perhaps suffer the most. Among the big cats they alone prefer to hunt by day and in the open, chasing down their prey in a 200-metre dash. Thus they are the only hunters that are easy to see in action, and people crowd around them to watch them at work. The clamour of minibuses in the cool of the morning and evening forces cheetahs to hunt in the middle of the day, which causes them enormous physiological stress. The circle of buses as they feed attracts hyaenas and lions, which easily drive them from their kill. Richard Leakey wants to put the whole operation on a sounder footing: fewer vehicles, but bigger; out-of-bounds areas; better roads (to avoid ploughing up the bush); more rangers, with better equipment.

Culling wild creatures for meat and milk, too, is far harder than it may seem. Animals intended for a general market should ideally be slaughtered humanely and prepared hygienically; there is a difference between a carcass and a clotted corpse. But if this requirement is taken to its logical conclusions – refrigerated mobile abattoirs, out in the bush –

the game becomes prohibitively expensive. On the other hand, if the animals are restrained (ranched) to make them more accessible, then some of their biological advantages are lost. For example, eland are wonderfully efficient producers of milk in arid conditions, provided they are allowed to forage far and wide at night. If they are coralled at night so that they are available for milking in the morning, and so are forced to feed by day, they begin to suffer the same distress as cattle.

To be sure, there are conflicts of approach, even among different shades of utilitarian. African elephants provide a neat illustration. Their numbers have diminished appallingly in recent decades: from an estimated 1.5 million in 1979, to between 500,000 and 750,000 today. To some extent, the decline has been inevitable; and with the best conservational will in the world it may be impossible to avoid further decline, as Africa's human population continues to expand more quickly, in places, than anywhere else on Earth. Culling has also been necessary locally, to avoid the devastation that is known to ensue when a reserve contains more elephants than it can sustain. But to some extent, the decline has been caused by poaching, which is both cruel and random; a slaughter, rather than a rational diminution of population that should be achieved by controlled culling. Poaching occurs because elephants have tusks, made of valuable ivory. Every African country could benefit from more income; every African country with elephants could derive at least some income from ivory; but everyone agrees that poaching is a bad thing (bad for elephants, and bad for the economy) and would like to stamp it out.

So – is it better for governments to sell the ivory that they acquire by legitimate culling, and use the money for further conservational purposes? Or should they refuse to trade in ivory and seek to destroy the world market, and thus take away the incentive to poach? Or is some middle course possible?

Each possibility has its advocates. Dr David Jones, director of London and Whipsnade zoos in England, argues that sale of ivory legitimately obtained is not only sensible, but positively desirable; precisely because such a sale provides local people with a reason to conserve elephants for themselves, and with an income that will help them to guard against poachers.[2] Richard Leakey argues that the only way to suppress ivory poaching is to undermine it: to make it unfashionable to own ivory, just as it is now unfashionable to own a fur from a spotted cat. This is why he supported the decision of the Kenyan government in 1989 to burn US$3 million's-worth of ivory; and (in similar vein) to burn a huge pile of rhino horns in 1990. He also argued[3] that Britain's decision to unblock the sale of 700 tonnes of ivory in Hong Kong led directly to an increase in poaching in the following months.

Dr John Beddington, at Imperial College, London, suggests a third strategy; though, he stresses, he does so dispassionately, and not as an advocate. Tusks, he points out, grow throughout the elephant's life. Furthermore, their growth is exponential; that is, the bigger the elephant gets, the faster the tusks grow. Thus, he says, it would pay a businessman to buy futures in ivory. He could borrow money at (say) 15 per cent; buy an elephant; and every year the elephant stayed alive his investment in it would grow faster and faster. This is a nice twist; animals would be exploited for their ivory by keeping them alive. We could envisage reserves where septuagenarian tuskers lived out their declining years, jealously guarded by the borrowed wealth of some far-flung millionaire.

I do not presume to say which of these strategies is right. Ethically, there is not much to choose: all are well intentioned, and the animals whose ivory might be sold are those that would probably be culled anyway, while a touch of Beddington logic could give the enterprise a more benign mien. All that matters, it seems to me, is that whatever is done should work; elephants should be kept in safe numbers; the control of populations should be as untraumatic as possible; and poaching should be eliminated. I am discussing the rival strategies only because they are interesting.

In principle, after all, all practical problems are there to be overcome. In general, these 'utilitarian' arguments must hold. It is necessary in conserving animals to try as far as possible to bring benefit to local people, for otherwise the conservation effort is doomed. Neither can we doubt that such policies can work. For example, the reintroduction of the Arabian oryx to Oman, which I will discuss throughout this book, has generated more income over the past ten years for the Harasis people who are acting as its wardens than the oil industry has done.

Yet there are larger objections to the utilitarian arguments, which lead me to believe that although they may be necessary, and must be acted upon, they are not, by themselves, sufficient.

HOW MUCH WILDLIFE DO WE REALLY NEED?

The first of the larger objections is that the most high-sounding of the utilitarian arguments simply is not true; or is not true enough, at least, to carry the day. Tropical forests (which contain most of the world's biodiversity; at least 90 per cent of all species) do not, in fact, make the greatest contribution to atmospheric chemistry. Unicellular marine algae, floating inconspicuously in the plankton, probably absorb 80 per cent of the CO_2 that other creatures produce.

More to the point, the ability of forests to absorb CO_2 derives not from their biodiversity, but from their biomass; and although the point is often made, the fact is that the diversity does not confer the stability that makes the mass possible. The diversity may appear to be important; but that is largely an illusion. Thus, it is certainly true that monocultures – areas of vegetation containing only one species, or only one variety of one species – are extremely vulnerable to epidemic or to change or climate; and Third World farmers who cannot afford pesticides are advised to grow many different crops, or to grow 'varieties' (or, more accurately, 'landraces') of crops that are extremely variable genetically. However, a forest would probably be reasonably safe from extermination by any particular kind of pest if it contained only a few dozen clones of a few dozen species. It would not need the thousands of tree species found in natural tropical forests worldwide, and still less would it need the millions of insects that accompany them. It is true, of course, that there are many interdependencies in natural tropical forest, so that if one species disappears then many others go by the board as well. But this shows only that natural forest is intricate and fragile. It does not demonstrate that total biomass *depends* upon diversity; for the remaining species, though fewer in number, grow to fill the gaps left by the absentees. In short, if we wanted tropical forest simply to mop up CO_2 and prevent soil erosion, then we might just as well plant a reasonably mixed plantation of commercial trees – a few dozen clones of rubber, eucalyptus, and so on. This would give us the biomass we required, with a cash return as well.

More generally, the utilitarian arguments are insufficeint by themselves because, if they are taken to their logical conclusion, they are self-defeating. After all, if we argue that wildlife should be kept because it provides local people with income from toursits, then we must ask: 'Should that wildlife be retained if the local people find a better way of earning money?' Thus, we can admit that many people in Kenya make a better living as game-wardens or as waiters, than they could do if they cleared away the wild animals and raised sorghum. But such an argument does not always apply. The mangroves in Florida's Everglades provide jobs for many local people (including local Indians), as tour guides, shopkeepers, and so on. But further up the coast, at Miami, the mangroves have been swept aside to make way for hotels and casinos, which provide thousands of times more income for thousands of times more people. What price the wildlife? And what price the wildlife in Lake Malawi, now that oil has been found beneath it? What price the wildlife in Serengeti, if diamonds and oil were discovered there? The problem is that the utilitarian argument is always temporary. Logically, wildlife can be defended as a source of income only so long as nothing more lucrative comes along. As soon as it does, then the logic of the argument demands that the wildlife be swept aside to make way for it.

The utilitarian argument leads us into another logical trap. Suppose we demonstrated beyond reasonable doubt that tourists really do offer Kenya its best source of income; and that the animals should therefore be preserved for ever. Still they would not be safe. For tourists are in a hurry. They do not come to see honey badgers and fennec foxes, which are creatures for the connoisseur (and which only a specialist would ever see). They quickly grow tired of antelopes. They want to see elephants and lions; and, preferably, they want to see a kill.

So if it's tourists we want, and the money they bring, why not give them what they are paying for? The lion population could be augmented with supplements of beef, the elephants with hay. Paths could be laid and waterholes artfully dug to give the tourists the greatest illusion of wilderness, and the greatest chance of happening upon a kill. Any animal that did not contribute to the coffers could simply be left to hazard. Wildlife managers already tread a difficult path, and only just (so far, and for the most part) avoid this trap. But if the money argument is allowed to prevail, then the trap will be seen, rather, as a goal.

Even the Norman Myers argument raises doubts – that unexplored creatures may yet provide us with drugs and crops. This may well be true; and it seems indeed that if there is a cure for AIDS, it may derive from agents ('sugar mimics') that derive from leguminous plants, and would interfere with the construction of the AIDS virus coat. Yet it will not be easy to scour the world's remaining plants for useful chemical agents; and not being easy, it will not usually be cost-effective. An alternative strategy these days is simply to work out what kind of molecule is liable to produce the required pharmacological effect, and then synthesise it. Besides, people who have to choose between the preservation of a forest that *may* yield a useful and profitable drug in a few decades' time, or clearing the forest for cattle to provide income now, are likely to say 'No contest'. They would say that if they were hungry; and they would say it if they were businessmen, seeking to maximise return on investment.

In practice, the Norman Myers argument is most powerful when stated simply in a general way. Thus we might say that we already have evidence that wild creatures *can* be good for us (*vide* the Madagascar periwinkle). From that, we can reasonably argue that biodiversity in general (not simply the variety of entire creatures, but the totality of genes that they contain) should be regarded *en bloc* as a resource, which we should pass on intact to our descendants; and we should do that, even though we may not know how to make use of the resource ourselves.

Even so, the ethics of accountancy that has prevailed through the 1980s would question this argument. Is such a vague and general 'resource' really more 'valuable' than the wealth that we might leave, if we spread a few more farms over the wilderness? Besides, these are still very early days for molecular biology, yet already we know how to synthesise genes. Why keep them 'on the shelf', in recondite plants and animals, when we could make them to order? And then again, it has often been argued in recent years, in many contexts, that posterity ought to be able to take care of itself. The ethical decision to leave our descendants a putative resource is, when you think about it, not so utilitarian as it first appears. Why should we be nice to people who will not be born for centuries to come? Are such people any more use to us than Sumatran tigers?

Neither should we forget – nature will not let us forget – that wild creatures were not created exclusively for our benefit; or if they were, they certainly do not act that way. It is commonly argued that only modern generations have 'lost touch' with nature, to the extent that they conspire in its destruction. This hardly seems to be the case. The people of the Middle Ages may have been 'in touch' with nature, but they did not exactly rejoice in this. They feared nature, at least as much as they admired it. Their gardens were knots and espaliers – an artful rejection of nature. Forest and sea were dark and threatening. The people of the 14th century did not know that the Black Death was caused by a bacterium carried by rats, and emanating from ground-dwelling rodents in Asia; but if they had, this would only have increased their conviction that much that lay beyond their city walls was the work of the Devil, and best put down. Paradise, as conceived by the ancient Persians, was a garden; not a natural place. So was Eden. Throughout the Bible, indeed, 'wilderness' was synonymous with threat and hardship. People feared wetlands until the 20th century – and still are over-keen on draining marshes – because they feared the ague, which they felt rose from its vapours. They were wrong in biological detail, but right in principle, because marshes harbour the mosquitoes that carry the parasites of malaria which, until recent decades, flourished in Europe as well as the tropics. The English Romantic poets and painters made much of wild nature, but in fact contrived their landscapes largely in the studio, and admired landscapes – the Lakes, the Scottish Highlands, Tuscany – that had been re-shaped by human endeavour. It is only the people of the late 20th century who have lost their fear of nature, because they know they can so easily obliterate what they do not like. It is only *us*, indeed, who can afford to talk so airily of 'biodiversity' and can seek – as far as is now possible – to conserve nature in its pristine state. Even now, though, we still seek (very reasonably) to eliminate the parasites of malaria and sleeping sickness, and the insects that carry them. Even now, people still fear wolves. We cannot be too literal, then, in claiming that what is good for other creatures should be good for us. Mosquitoes and the parasites they carry are not.

Suppose, then, that we took the utilitarian approach to its logical conclusion; and suppose we knew a great deal more about biology than we do at present. Then we might reasonably appoint a panel of biologists – similar to the SCOPE committees that now report to the world in general on problems of the environment – to draw up a list of creatures that really were of benefit to humankind, and should be kept for that reason. Suppose we interpreted 'benefit' liberally: to include not only those animals, plants, fungi and microbes that provide us with food and traction and drugs, but also those that we like to look at, and fondle; and those that would help to preserve biological stability, and keep the atmospheric gases in check. The list would include, in short, the Congo peacock as well as the chicken, the okapi as well as the Friesian cow. Yet such a list (I suggest) would be unlikely to include more than about 100,000 species. That would be far more than most professional biologists would ever get round to knowing, and probably a thousand times more than most people would recognise. But if we created such a world, with happy and secure human beings and a list of other creatures that would extend beyond the dreams of almost everybody, we would still have perpetrated a mass extinction. One hundred thousand species is less than a third of 1 per cent of all the creatures that now are thought to live on Earth.

The utilitarian arguments are powerful, then. Indeed they are inescapable; we must find ways of reconciling conservation with human well-being. Yet they are insufficient. By themselves they do not seem to provide convincing reasons for conserving more than a tiny minority of the creatures that now exist. Indeed they are circular; they provide reasons for conserving only those creatures that can be seen to bring us direct benefit. So what else can we say in answer to the question, 'Why conserve animals?'

WHAT OTHER REASONS ARE THERE FOR CONSERVING ANIMALS?

When oil is discovered beneath some wildlife reserve, it is no longer enough to argue that the wildlife is worth preserving because it brings an economic return. It may be possible to argue that with a 100-year projection, for oil-fields run dry while ecosystems can persist forever. But as the adage goes: people eat in the short term. When this happens, then, and it will happen more and more, the only argument left is the ethical one: that the animals should be conserved because it is right to conserve them. Of course they should probably bring some economic return; but it is the ethical point – that their conservation is 'good' – that will enable them to prevail even when their destruction could bring even greater return. The economic return from tourism, in short, should not be seen as the reason for conserving animals. Tourism merely makes it economically possible to do what is right.

How, though, can we defend this ethical position? What arguments can we bring in addition to the utilitarian, to show that any particular course of action is 'good?'

The written evidence suggests that philosophers have been asking that general question at least since the authors of *Exodus*; and probably (for how could it be otherwise?) for tens of thousands of years before that. There are university departments devoted to ethics, and entire libraries of learned treatises. Arguably the questions 'What is good, and why, and how do we know what is good?' are the principal cerebral preoccupations of humankind. Yet all we can reasonably conclude is that different philosophers have come up with different kinds of answer. Some have indeed been content simply to be utilitarian; to argue that what is good, in the end, is what is good for human beings, or at least for the maximum number. Some appeal to God – that what He says is what is right; though there is little agreement on what it is He does say, or which of the world's several versions of God should be seen as the authority. Some, like Immanuel Kant, have sought fundamental principles separate from God and separate from mere expediency on which to found an unassailable ethical structure, to which even God should be subject. Neither Kant, nor anyone else, has succeeded in this; or only, at least, to the satisfaction of themselves and of their followers.

I continue to be absorbed by moral philosophy, and do not consider it an idle pursuit. It is possible to discover at least some principles that seem to give more weight to some arguments than to others. It is certainly reasonable to follow the example of individual human beings – Jesus Christ, Mahatma Gandhi, the Dalai Lama – who seem to us to be 'good'. But so far, no one has succeeded in answering unequivocally the questions, 'What is good, why, and how do we know?' David Hume suggested in the 18th century that we must in the end be guided by our 'feelings'. Arguments can be brought to bear to defend feelings – and this, in the end, is what moral philosophers do. Attempts can be made to cultivate finer feelings – which, I suggest, is the role of religion, with prayer and contemplation being part of its armamentarium. But in the end, what is right is what we feel in our bones to be right. Hume did point out, however, that there is a remarkable degree of correspondence between the bone-feelings of different people: the injunctions that we should not kill, or behave too selfishly do have a consensus that runs through many centuries and societies – so bone-feelings of 'rightness' are not as quirky or as individualistic as might be feared.

For my part, though, I have junked all the hundreds of pages in which I have attempted to explain why I feel in my bones that it is right to conserve animals. I cannot significantly improve on the assertion that it simply is proper for us, as intelligent members of the universe, to try to look after our fellow creatures, and evil for us to do otherwise.

But if I cannot find reasons worth putting on paper to defend my bone-feeling that it is right to conserve animals, I do think it worthwhile to discuss the implications and the consequences of such a feeling.

FEELINGS AND ATTITUDES

Underneath all our apparently rational decisions – to vote for this party or that; to create a wildlife reserve or sweep the wilderness aside – lies a conception of the world and of our place in it that is summarised in the word 'attitude'. Attitude is not an irrational thing, for it is amenable to rational argument. But it is in the end 'non-rational'. It is the summary of our bone-feelings about what is good and what is bad.

Each of us is capable of adopting many different attitudes. As the Old Testament says: human beings have choice, and the fundamental choice is what 'attitude' we choose to adopt. Societies have prevailing attitudes too, in so far as they sometimes allow one kind of personal attitude to prevail, and sometimes another.

Historically, and prehistorically, human beings have regarded the world about them and the creatures therein with one or other – or generally a mixture – of two main attitudes. At times they have regarded the rest of the world as a resource, to be explored and to be exploited. At times they have regarded the rest of the world with reverence, and treated it conservatively. The poetic and theological way to express this latter idea is that the world as a whole is a Creation, and we are part of that Creation. This may be literally the case, or it may not; the idea may simply be a metaphor. What is important in either case is that the metaphor summarises an attitude; and attitudes are difficult to summarise without recourse to metaphor.

Each of these principal attitudes to the world – exploitative–exploratory; or reverential–conservative – has its own advantages and disadvantages. In general, the conservative attitude is 'safe'. Each generation of human beings observes what the previous generation did, and follows suit. If the previous generation succeeded, then the chances are that the one that follows will do so too. If the previous generation failed – well, there would be no succeeding generation. Archaeological and palaeontological evidence suggests that human beings have at times been extremely conservative. Thus *Homo erectus*, the species of human who preceded *Homo sapiens*, was apparently doing the same kinds of things 1 million years ago that he and she were doing 2 million years ago. The interval in between is known to palaeo-anthropologists as 'the million years of boredom'. But the strategy succeeded. What worked for the early *H. erectus* worked for the late ones too.

To be exploratory and exploitative is riskier. The explorers may simply do something wrong, and then they will die out. If they exploit too vigorously, then they will kill their own resource, and again die out.

Yet when the exploratory–exploitative mode succeeds, it can succeed beyond the dreams of those who are reverential–conservative. And 'success' here is employed in its Darwinian sense: the successful explorers–exploiters will leave more offspring than the reverential–conservatives. If the two kinds ever find themselves in opposition – as rivals for the same territory, or herd of mammoths – then the successful explorers–exploiters will prevail. Thus, over time, risky though it is, natural selection tends to favour the exploratory–exploitative mode. In the long term, to be sure, the explorers–exploiters may do worse than the reverential–conservatives. But in the short term they push the reverential–conservatives aside. And once any group is gone, it is gone.

I used to argue that there was a clear cut-off between reverential–conservative and exploratory–exploitative which coincided with the rise of farming, around 10,000 years ago. After all, the myth of the 1960s (which is when I first started thinking about these things) was that hunter-gatherers – pre-farming people – were 'in harmony' with nature; while farmers tended simply to impose their will on nature.

This, I now feel, is naive. The archaeological record shows many examples of hunter-gathering people who were extremely bullish in their exploitation of nature. The Maoris – Polynesians who colonised New Zealand after the 9th century AD – wiped out most of the 25 or so species of moa that used to prevail there (leaving Europeans to help finish off the rest). The hunter-gathering civilisations of Easter Island wiped their environment clean before dying out themselves, about the time of Christ. It still seems possible that the first people to colonise America across the land-bridge from Asia (the land-bridge that is now the Bering Strait) destroyed all the largest animals as they migrated south – elephants, camels, ground sloths and the sabre-toothed cats that had preyed upon them; a putative wipe-out known as the 'Pleistocene Overkill'. To come up to date, modern fishermen tend to be extremely unconservative; while whalers have to be almost physically constrained to prevent them from driving their quarry to extinction.

In short, it just is not true that hunters are conservative. It is just that those who do manage to exercise some restraint are the most likely to survive in the long term; so that the hunting peoples who remain in the world do seem to show a reverential attitude to the creatures they kill, and to nature as a whole. The Australian aborigines perhaps epitomise the trend. They in fact exploit their environment extremely efficiently, notably by the use of fire to freshen up the vegetation. But they came to the island continent from Asia at least 40,000 years ago – some say 80,000 or more – and although they may well have helped to drive many of the larger animals to extinction when they first arrived (including giant marsupials and some enormous reptiles) they seem to have lived in harmony with what remained at least for the past 10,000 years. The hunter-gathering aborigines whom the Europeans first encountered still killed animals, but they also revered them. They still use the land, but they feel that they belong to it, not that it belongs to them; they have no sense of land ownership. Arguably, they have succeeded better than any other people that have ever existed in striking a balance between exploitativeness, and conservative reverence.

But farmers too can be conservative in mood, and reverential in attitude. Traditional farmers commonly felt that it was wrong to take out every tree, and extend over every hillside. Harvest was a time for thanksgiving: in modern societies, generally to God; but in more ancient societies, often to the Earth in general for its generosity. We cannot then discern a clear difference in attitude between hunters and farmers; and sometimes, indeed, farmers seem the more reverential, and the more conservative.

Yet hunting has built-in constraints that farming does not. If hunters are too vigorous, they wipe out their resource; and many a hunting society has done just that. But if one farmer is more vigorous than his neighbour he may succeed in growing more. The difference is that the whole point of farming is to *increase* the resource. Sooner or later a farmer will reach a limit; he can over-reach himself just as the hunter can. But that point lies far off. Until he reaches it, the rule applies that the harder he works, the more he will achieve. At a purely technical level, it is clear that farming is innately exploitative, for to sweep one plant aside and plant another is to take nature by the scruff. More to the point, farming is exploitative because it rewards exploitation. Hunters all too easily over-hunt, as modern fishermen and whalers are busy demonstrating. But in farming, in general, the more effort you put in, the greater the reward.

Thus, I no longer believe that there was, in reality, a simple shift of attitude between the hunter-gathering age of the Upper Palaeolithic, and the farming of the Neolithic. I do believe, however, that once farming was in train exploitativeness began to pay off more consistently than it could have done before. However reverential and gentle we may aspire to be today, we are, inescapably, the inheritors of an exploitative way of life.

In the past, philosophers and biologists tended to take a romantic view of evolution. They supposed that evolution by means of natural selection necessarily led to forms or ways of life that were better than what had gone before. There was little excuse for this. Charles Darwin, who first properly formulated the idea of evolution by means of natural selection, emphasised that it did not lead necessarily or inexorably to improvement.

He merely said (to put the case in its weakest form) that what succeeds is, well, what succeeds.

However, philosophers and anthropologists have commonly applied this idea, that evolution means improvement, to the advent of farming. They have often assumed that the life of the hunter was, as Thomas Hobbes said of human life in general, '. . . poor, nasty, brutish, and short'. Farming (the traditional view had it) came as a Godsend (literally). People no longer had to run the risk of the wilderness. They could simply cultivate their own gardens; a way of life that was not only plenteous (the supposition had it) but also comfortable.

Recent archaeological and anthropological research suggests that the truth is exactly opposite. Hunter-gatherers who live in favourable places (as some Australian aboriginals do even today on Cape York) probably had an extremely easy life. Migrating birds turned up on cue; there might be shellfish for the taking; there was always some tree in fruit. Farmers, by contrast, lived a life of horrendous toil, forever likely to be struck down by drought or by locusts (as indeed is graphically recounted in *Exodus*). Indeed, the myth of the Garden of Eden (in *Genesis*) can reasonably be seen as folk-memory of the end of blissful hunter-gathering and the beginning of agriculture. God's words to Adam as He banishes him from the Garden are: 'In the sweat of thy face shall thou eat bread!' And indeed Adam must have sweated. The remains of neolithic skeletons show appalling signs of over-use, as people ground the corn with stones; and also of malnutrition, as the varied diet of the hunter-gatherer gave way to one based on cereal.

So why did the Upper Palaeolithic/Mesolithic hunters abandon their easy way of life in favour of farming? Because of natural selection – natural selection in the sense that Charles Darwin himself envisaged, not in the sense that some later writers have interpreted it. Hunter-gathering can support only a limited number of people; as many as the natural environment will provide for. Farming is designed to increase the output, and hence increases the number who can be supported. John Yellen argues that the true origins of farming probably lie 20,000 years or so before it became sufficiently widespread to show up in the fossil record. Throughout that time, people probably combined a little cultivation – the planting of favourite trees; the damming of streams, perhaps – with more traditional hunting. Modern Hottentot people in the Kalahari desert still do this; running a few goats in some years, then getting rid of them and reverting to hunting again.

However, the people who pursued this agreeable way of life were bound to multiply more than those who were simply hunters, precisely because they were producing more food. The more they multiplied, the more it became necessary to cultivate. Probably – undoubtedly – many would have decided that to cultivate more and to multiply was less agreeable than to cultivate less and keep the population small. But this is where natural selection came in. Those groups who did not opt for the agreeable way of life; those who decided to cultivate more – inevitably ousted those who preferred to stay the same. Many thousands of years later, we can see a form of that clash in Australia, as the hunter-gathering Aborigines, stable and secure though they were, were swept aside by the more vigorous and multiplicative Europeans with their long history of farming. In short, farming is not *better*. People were not waiting for it to happen, with bated breath. Almost certainly, for many thousands of years, they resisted it. But it was bound to succeed in the end, because it can support more people.

And it was thus, by the seemingly inevitable rise of agriculture, that the exploitative mode was vindicated. It did not suddenly come into being, but eventually it came into its own. The farmers of classical times, of Rome, Greece and Egypt; those of the Middle Ages, worldwide; and many 'traditional' farmers today demonstrate how adept the *craft* of farming can be. Even the most modern of modern farming is still largely shaped by the practices that were developed thousands of years ago. But even so, farmers until even beyond the Middle Ages were severely limited in what they could achieve. They could not afford to experiment – explore – too much, for any deviation from tradition was more likely than not to lead to disaster.

The coming of science to agriculture from the 18th century onwards changed all that. With science, it is possible to experiment formally; to demonstrate unequivocally that some innovation works, or does not; and in the end, to show why it works, or does not. With such an approach, farming could at last begin to fulfil its promise; to push the output of the world not simply to its ecological limits, but to its physical limits. There have been several reasons why the human population has risen so spectacularly these past 200 years, and continues to rise, including the contribution of medicine. But it is the broadening and deepening of the food-base that is the *sine qua non*; and that has depended on the partnership of agriculture and science. This partnership has finally vindicated the exploratory–exploitative way of life. The explorers–exploiters have swept all before them: all of nature, and many of their more conservative fellow humans. In the past 10,000 years it has paid, on balance, to be exploratory–exploitative. In the past 200 years it has paid a hundred times over.

But as we will see in the next chapter, the party is now over. We seem to have pushed the world almost as far as it can go. The fabric itself is declining, faster than it can be restored. It is clear that some time in the next few decades we will have to bring the growth of the human population to a halt, or else it will simply crash. In short, the exploratory–exploitative mode, by itself, is no longer appropriate. We must relearn the habits of conservationism.

Note, to anticipate a later point, that I absolutely do *not* advocate a simple return to some earlier way of life, as some of my hippie contemporaries from the 1960s were wont to do. As a matter of fact, I do not believe there ever was a way of life or an attitude that would exactly suit the needs of today; and even if there were, it simply is not an option to turn our backs on modern science. Indeed this book will show that science, well deployed, is vital if we are to rescue what is left of the planet and our fellow creatures. Science can, though, just as soon be applied to a reverential–conservative way of life as it can to an exploitative one. Indeed, the tradition of reverence is as powerful in science as is that of exploitation. Isaac Newton, among the founders of modern science in the 17th century, was primarily interested in theology; and he (like many of his contemporaries) regarded science not as a means by which to change the world, but as an insight into God's purpose. For him, scientific exploration was itself an act of reverence. But in the past few centuries we seem to have put that tradition to one side, just as we have tended to put reverence in general to one side.

The immediate question is *how*, after all these centuries and millenniums, do we return to a more reverential approach?

MYTHS AND TABOOS; REVERENCE AND RELIGION

There are various reasons why it is difficult simply to restore the habit of reverence. One is that many people simply do not see the point of it; only a minority, indeed, are likely to read books such as this which point out the necessity. A second is that it is difficult to find arguments sufficiently brief, and with enough rhetorical power, to persuade people whose minds are on other things. The deepest problem of all, though, is the one that Christ was constantly confronted with. For the fact is that in the short term, even now, exploratory–exploitativeness succeeds. Even today, a nation that sets out to be bullish (by whatever means) can crush one that does not. The reverential–conservative way of life will bring rewards in the long term: a more agreeable and stable way of life. But in the short term, the fight is to the aggressive.

Christ argued his way out of this dilemma with adages: 'Blessed are the peacemakers, for they shall inherit the world.' More generally, though, anthropology suggests that human beings, from the time that they first became conscious, overcame the temptation always to opt for short-term benefit, by inventing or evolving myths and taboos. Thus as Marvin

Harris of the university of Florida has argued in books such as *Good to Eat*,[4] the Jews 'invented' a taboo against pork (described in *Leviticus*) not because pork is, literally, nasty, but because it is all too attractive. The real trouble was that any attempt to raise pigs in the desert where the ancient Jews were obliged to live would have been economically disastrous, because pigs need shade and water, and a mixed, high-energy and protein diet. The only suitable livestock, though probably far less tasty, were drought-resistant sheep and goats that could subsist on thistles. Similarly, societies that have relied on horses for transport and warfare have learnt to be disgusted at the thought of eating them. Otherwise they would destroy their own livelihood whenever they became hungry. The Hindu sacred cow is sacred because it is *necessary*. The cow is the mother of the bullocks that pull the ploughs; the root of all the economy.

It is possible that such myths and taboos were devised consciously by wise old prophets, who could see far ahead, and knew that their people needed stories to cling to. It is equally possible that such myths evolved by natural selection. People who had such a view of the world survived better than those who did not. Either way, the fact is inescapable that societies that have succeeded have done so largely by adhering to deep and unexamined beliefs – that such and such a course of action is bad, and such and such is good – even though there may be immediate advantage in ignoring those beliefs.

Ancient Jews and Hindus lived a long time ago, by definition. But they were not 'primitive'. They were no different from us. The technique they adopted to keep them on the straight and narrow is entirely appropriate to ourselves. If we want to begin to treat the world reverentially and conservatively, and in particular if we want to conserve animals, then we have to create and adhere to some myth; some version of the ancient notion that animals are divine, and that to destroy them is a sin. The point is not so much to argue the case as to find a persuasive way to express it. Michael Soulé likes to quote Albert Schweitzer's notion that animals and plants are our brothers and sisters, and that we, as the most competent members of the family, are obliged to look after them. It is as good a way to state the matter as any.

Note, though, that the development of myths as guides to action *is* religion. Conscious thinking and argument can influence the kinds of myths we care to create. But the processes by which those myths are created are those of contemplation, and shared experience. Aborigines would have no trouble with this. But we sophisticated Westerners, who have the world in our hands, and have so comprehensively demonstrated these past few hundred years that an exploitative temperament guided by 'rational' thought is the key to success, have to relearn ancient habits. Whether the traditional religions can rise to the task remains to be seen.

Clearly I am suggesting that unless we become more conservative, we will run into very deep trouble; overstretch the fabric of the planet, and destroy the base of our own existence. This surely is undeniable. Yet there may seem to be some conflict between this idea and the notion stated earlier; that there is no direct relationship between ecological stability and biodiversity. There is no conflict, however. Whatever we do, we need to treat the world more conservatively if we are to surive. We cannot continue to breed as quickly as we have been doing. But there are, in theory, many different ways of creating ecological stability. One way is to try to conserve as much as possible of what now exists – including as many as possible of the remaining animals. The other is to draw up the list of species we would theoretically need to keep the planet ticking along, and the ones that obviously bring us benefit, and conserve only them; letting the rest go to the wall.

The latter course – being conservative, but only in so far as this obviously benefits us – is what I would call 'environmentalist'. It is the kind of idea that underlies most political speeches on 'the environment'. It is clear that most politicians do not know the difference between a wildlife reserve, created to preserve biological diversity, and a golf course, which may look nice, and give a few people some pleasure, but might contain very few species.

The more expansive approach – seeking to conserve as much as possible of present biological diversity – is the philosophy that in Europe is called 'Green'. Green philosophers do not argue that we should simply create a world for our own comfort, and forget every creature that does not seem to be of use. Greens argue as I have argued above, that other creatures should be valued for their own sake. This philosophy indeed implies some discomfort for human beings. After all, we may well ask people to give up dreams of a marina in favour of a mangrove forest, even though you have to have an inordinate fondness for spiders (as I can attest) to appreciate mangrove forests. Green philosophy is to this extent 'perverse': it invites human beings to pursue a course that is not in their own direct material interest. It is because of this that the appeal to ethics is necessary: pursuing a course of action simply because it is perceived to be 'good'. And it is because such courses of action are fragile that they need to be supported by carefully cultivated attitudes, and protected by religious myth.

Of course we must concede that Green policies are most likely to succeed when they provide some pay-off; when they provide local people with an income. That is why it is essential to contrive some pay-off. But the pay-off is not the reason for being Green. It merely provides the excuse. By the same token, Green philosophies should produce a more stable world than the one we are contriving at present. But that is not the reason for being Green. That is merely the bonus. By the same token, Christ did not suggest that Heaven was the reason for doing good. It, too, was a bonus. And there are, of course, 'low-church' Christians who argue that Heaven was never meant to be a literal place; but was, rather, a metaphor for a way of life and a state of mind that could be achieved on Earth. Such an idea fits Green philosophy very well.

One further digression is necessary. I have said that a Green world, which contains creatures that are not necessarily of great benefit to us, is not the same as an environmentalist world, which is tailored for our own predilection. It follows, then, that in creating a Green world we would be making some sacrifice – because, by so doing, we would eschew the possibility of creating a world that would meet our material needs precisely. How much sacrifice is it reasonable to make?

HOW MUCH SHOULD WE SACRIFICE?

Some people who are eminent in conservation circles (politically rather than biologically) have been known to say that they prefer animals to people. Sentiments like that get conservation a bad name; they confirm in the minds of those who prefer to do nothing the notion that conservation itself is bad, because it conceals an extremist form of anti-humanitarianism. But I know no serious conservation biologist who is anti-humanitarian. The general argument is that human beings *must* curb their own population growth, and eventually reduce their numbers (by benign and voluntary means) in their own self-interest; and that other creatures can flourish, and should be encouraged to do so, as this diminution takes place.

Of course, there do sometimes seem to be direct conflicts between the needs of animals and those of humans. Livestock farmers do lose stock to wolves, tigers, and cheetahs (which can make short work of a flock of sheep, just as foxes make short work of a run of chickens). Arable farmers and planters do lose out to cockatoos, monkeys, and elephants. Such conflicts, however, are small-scale and man-made. If the world took conservation even slightly seriously, then it would be possible for us to buy our way out of such problems; partly by paying compensation, and partly by building fences and ha-has to confine the animals.

The larger conflicts occur when farmers spread into land previously occupied by wild animals: goats into the Sahel, once the domain of Scimitar-horned oryx; cattle into reserves throughout Africa and Asia. It seems extremely high-handed for Europeans to suggest that farmers should not spread out when they need to. We, after all, have long since taken over most of our continent's fertile valleys, and pushed aside most of the deciduous forest. It also seems inhumane to tell Sahelian farmers they should not try to make a living from the desert – for there is nowhere else for them to go. If the oryx die – well; so did the boar in England, when we cut down our woods.

European objections are not simply self-righteous, however. The fact is that rich, temperate lands, well farmed, can support enormous populations, relatively cheaply. Desert is not good farming land, and could not become so unless huge amounts of money were poured into it. Farmers in the Sahel make a very poor living, with yields of grain (sorghum), or of milk and meat, which are sometimes only a few per cent of what can be achieved in Europe. It is a bad use of land, by any standards, and the people and their livestock suffer from its inappro- priateness. They are not there by choice, however, or because they are foolish. They have been forced into this unpromising territory. The reason they are forced is not, in general, because the good agricultural land in their own countries is too scarce to accommodate them, but because it is owned by the rich, and is not available to subsistence farmers. People who farm unpromising land are, in general and in effect, exiles in their own countries. But it is politically easier (because it is cheaper, and does not incommode rich landowners) to push people into the wilderness than to make it possible for them to stay on richer lands. The 'conflict' between people and animals is, in such common cases, a sham. The real conflict is between poor people and rich people. The rich people make no concessions; so the poor are forced to do metaphorical battle with wild animals.

Again, if the world at large truly cared then we would do something about such conflicts. It would not cost very much (compared with the cost of most of the West's indulgences) to pay Sahelian farmers not to farm, but to look after oryx instead. Indeed it would cost a great deal less to do this than we spend on compensating the farmers of Europe or the US for not growing wheat or potatoes.

Yet we must concede that a Green world cannot be as comfortable for human beings as an environmentalist world, for the latter, by definition, is designed expressly to serve human needs. So how much discomfort should we be prepared to put up with?

Just to stick to generalities, it does seem absurd to ask people coolly to give up their lives to save animals. They just would not do this, even if it were 'right' to do so. But on the other hand, we cannot all aspire to live like the Queen of Sheba, or the average inhabitant of Hollywood. As Paul Ehrlich of Stanford University has commented, the average well- planned and self-satisfied middle-class Los Angeles family of two parents and two children consumes more than the average 'over-populated' Bangladeshi village. In the 1980s in particular, under Ronald Reagan and Margaret Thatcher, we were all encouraged to pursue personal wealth as a moral as well as an economic imperative; and there can be no doubt that this is incompatible with a Green world. If we want other species to survive in more than minimal numbers, then we must give up our personal dreams of untold wealth forever. That is the essence of the required sacrifice. The dream of untold wealth belongs among those of neolithic farmers, who lived at a time when the world itself seemed infinite. But the dream has become an anachronism.

We must accept, too, that although human beings should not be asked to die on behalf of animals, they must be asked to run some risk. Risk, indeed, is the sharp end of inconvenience. This should not be so difficult. We accept that risk is a part of everyday living. We accept that roads exact a toll; indeed a horrendous one. Every major civil engineering works, every bridge, tunnel, and dam, is almost *bound* to kill some of the workers. Yet we carry on building. People have been killed in the buffer zones around the reserves built for tigers in India as part of Project Tiger;[5] and some have argued that the Project should be discontinued because of this. But this in effect is to suggest that the species *Panthera tigris* is less important than a salesman's 'right' to drive his car at 70 mph. Fast driving causes a lot more deaths than tigers do. Even in Third World countries road accidents are high among the causes of death. Yet driving is seen as an 'acceptable' risk.

But of course we have to accept that the present world is crowded, and that human beings have rights as well. I have argued that we must make concessions for animals, and indeed be prepared to take risks. But I would not ask anyone to risk their lives to save, say, the parasite of malaria, or the particular mosquitoes that carry it. The issue is not quite as simple as it might seem. For example, there can be little doubt that the wild animals of Africa have survived as long as they have partly *because* Africa is overrun with tsetse fly, which carry parasites of genus *Trypanosoma*, which cause sleeping sickness in humans and severe disease in cattle and horses. Tsetses and trypanosomes have protected the wildlife by keeping humans and their cattle at bay. In a sophisticated world, though, we ought to be able to suppress the tsetse and still preserve the animals, simply because we want to. In short, being a Green does not imply being stupid. Obvious pathogens and their vectors that would upset the best-laid plans are best got shot of.

One further ethical question remains. This book is based on the premise that if we want to save the world's remaining animals, we must take their affairs in hand; that we must be prepared to curtail their freedom; that we must orchestrate their reproduction; that there will be times when we must kill individuals for the sake of populations, including populations that are yet unborn. This may all seem very arrogant. Do we have a right to behave so high-handedly?

SHOULD WE INTERFERE IN THE LIVES OF ANIMALS?

Jesus Christ was a pragmatist. He was always prepared to do what was necessary to achieve some greater good. The point is made clear in Luke 6:9. Christ has cured a man with a withered hand on the sabbath, and the Pharisees ask if it can be lawful to do such a thing on such a day. He replies, 'Is it lawful on the sabbath days to do good, or to do evil? To save life, or to destroy it?' That will do as a precedent for taking the affairs of animals in hand, even though we thereby impose on them. It is very difficult in this world – even for Christ – to do anything that is unequivocally good. But we have to do the best we can. And it does seem better to attempt to save animals from extinction than to let them die out; and if some imposition is necessary, well, so be it.

Neither does the notion that nature must be revered imply that we should therefore leave it alone. Many wild species have already been driven to such a parlous state that they cannot survive unless we take their affairs firmly in hand. Wild habitats that have been reduced in size (from the size of continents to the size of reserves) have to be managed or they are bound to degenerate. Caring, then, implies action. Specifically, it implies the application of very good science.

There are limits, though, to what we should allow ourselves to do. It is not permissible to be cruel to animals, to cause them avoidable pain and distress. In our efforts to conserve species, we always have to have regard for the welfare of individuals. But the techniques of modern conservation, which include contraception and arranged matings, are not innately cruel; and when it is necessary to cull – a quick and humane death is not the worst fate that an individual can endure, human or animal. A few individuals must be killed now, so that many thousands of others can continue into the future. Many a soldier has died for less, and with no more say in the matter. It is not 'good', necessarily, simply to shy away from what is distasteful.

In short, having made the decision to try to conserve as many of today's animals as possible we should, I believe, though without resorting to cruelty, do what is necessary. Not to do so is to be guilty of a sin of omission; and this is no less serious than a straightforward misdemeanour. One further touch of arrogance is necessary, though, if conservation is to succeed at all. We cannot save everything, and if we try, and spread our efforts too thinly, then everything will die. This is true in captive breeding, for as we will see in Chapter 2, zoos can save only a minority of the millions of species that are now threatened. It is true, too, of habitat protection, for it is not possible, as things are, to save more than a proportion even of the depleted wilderness that is left to us. So we have to choose: we have to be prepared to say, 'This species we will conserve, and this we will leave to hazard.' The criteria that modern conservation breeders employ in making this choice are discussed in Chapter 4. Perhaps it is arrogant to make these choices; but it seems better to risk the charge of arrogance and save some animals, than to safeguard our humility and allow everything to die. It is not virtuous to be effete.

So: the reason for conserving animals is that it is good to do so. Economic returns make it possible to do the things that are good, but they are not themselves the reason for doing good. The survival of other creatures must not depend upon our own immediate perception of benefit. Indeed in order to do good we must be prepared to make concessions, and suffer at least some inconvenience.

In order to be sure of doing good even when the benefits are not obvious, we must do as people have always done in those circumstances: cultivate taboos, metaphors and myths: ideas and forms of words to cling to, to carry us through. More generally, we must cultivate an attitude of reverence towards the world, rather than one of exploitativeness. The methods by which we may cultivate such an attitude include those of religion. Indeed, whether or not we care to acknowledge the existence of a God, or choose to adopt the idea of God as a metaphor, the cultivation of attitude *is* religion. A reverential attitude to nature, however, must not at this stage imply that we should simply let nature take its course. If we do that, then the world's remaining wildlife will continue to decline, for we have already rendered its existence precarious. It is ethically necessary to take matters in hand; and to do that we must apply good science.

The rest of this book is about the practicalities of conservation, and in particular about the contribution of captive breeding, especially as carried out in the world's zoos. But first we should define the scope of the problem, and ask what it is that zoos might in practice contribute.

TWO

THE SCOPE OF THE PROBLEM

In chapter 1 I argued that it is justified to keep animals in captivity for the ultimate benefit of their species – provided their short-term welfare is also attended to. 'Species', after all, are not simply the abstractions that non-biologists suppose. A species is a number of individuals, some alive, some already dead, some not yet born. 'Saving the species' means making it possible for creatures of the kind known as Arabian oryx or black rhinoceros to live in a hundred years' time, or a thousand – and perhaps a million (although, in the past 65 million years, the average longevity of each species has been only 500,000 years, by which time they either go extinct or – which is quite different – evolve into something else!). The basic moral position is that of pragmatism. It may not be ideal to deprive animals of their freedom. But it is rarely open to us to do unequivocal good and we generally have to settle for the least bad. I could have added that I would just as soon be a giraffe in San Diego Wildlife Park, say, as in Africa; that life in the wild is far from 'free' and I personally would trade theoretical liberty for an absence of lions and a depletion of parasites. But that might have seemed frivolous.

Yet there are critics who accept this moral position but still ask, 'What's the point?' After all, they argue, there may at present be millions of creatures in danger of extinction. Captive breeding can so far be credited unequivocally with saving only a handful: probably less than 20. Only a few score are at present being bred convincingly in captivity, to the point where we can say with reasonable confidence that we can save them, even if they do go extinct in the wild. At best, zoos as now constituted could probably save only a few hundred species from extinction; and the most optimistic zoo people predict at best a few thousand. On arithmetical grounds alone, then, the effort seems pointless – not to say pernicious, because it detracts from more serious endeavour.

There is a further set of arguments, too: that even if we do save animals in a zoo, they would be so hopelessly compromised by captivity that they could not return to the wild. Even if we could save the animal from actual physical extinction (this argument runs), we would achieve nothing worthwhile. We would simply turn a wild animal into another domesticate.

I will discuss this second set of arguments in later chapters. For now I want to address the arithmetic. What is the point of trying to save just a few hundred species, when millions are endangered? But we should start from further back. What is the basis of the critics' arguments? Why should we suppose that 'millions' are in danger? Surely that is a gross exaggeration?

Unfortunately, it is not.

HOW MANY SPECIES ARE THERE?

Biologists so far have described and named between 1 and 2 million species of creature. We will return later to the precise meaning of 'species', because in conservation the precise meaning is proving to be extremely important. But a good working definition is that a species is a group of creatures whose members are able to interbreed by sexual means to produce offspring that are fully viable, where 'viable' means capable of doing all the things that living things are supposed to do. Thus Alsatian dogs and poodles are the same species (*Canis familiaris*) because although they look very different, they are happy to mate and there is nothing wrong with their pups. But horses (*Equus caballus*) and asses (*Equus africanus*) are different species, because although they may crossbreed their hybrid offspring – mules – cannot be said to be 'fully viable', because they are sexually sterile. As we will see, life can be far more complicated than this, for there is nothing much wrong with the hybrid offspring of dogs (*C. familiaris*) and wolves (*C. lupus* or *C. rufus*) or coyotes (*C. latrans*) although they are considered to be different species. But I will come to such matters. The broad definition holds well enough.

'One to 2 million species' seems a little vague. Surely it is possible to say precisely how many have been named? It is not as easy as you might think. Human beings in general esteem the works of humans more than the works of God. Robert May, Royal Society Professor of Zoology at Oxford University, has lamented that you may find a central catalogue of the paintings of Rembrandt or of the books in the Smithsonian library, but no central list of the species with whom we share this planet.[1] One to 2 million does seem a lot, however. Furthermore, people have been recording the names of species at least since the times of the Greeks (though they undoubtedly started naming them many thousands of years before that); the modern system of classification was initiated 200 years ago, by Carl Linnaeus of Sweden; and since the late 18th century especially, expeditions have scoured the world in search of new life forms. So you might suppose that by now, biologists must have a pretty good idea of what is out there. Biologists are, however, a modest lot, and most admit that there are, in reality, probably between 5 and 10 million species. Recent preliminary studies in the tropical forest of Panama suggest, however, that biologists are not nearly modest enough.

It has been clear for some centuries that tropical forest contains many times more species than any other kind of habitat; Charles Darwin was especially impressed by the diversity of South American forest during his round-the-world voyage on *HMS Beagle* in the 1830s. The second most species-rich habitat is coral reef; but even that is many times less various than tropical forest. A temperate forest typically may contain only half-a-dozen species of tree, and sometimes (as in the Californian redwood forests) effectively only one. But the tropics as a whole contain around 50,000 different species of tree. Any one forest is likely to contain hundreds of species. And every one of those hundreds in turn supports hundreds of others: insects, mites, birds, mammals, frogs, fungi, epiphytes, and all the parasites that live within them.

In the late 1970s Terry Erwin of the Smithsonian Institute in Washington, DC set out to assess just how varied tropical forest really is.[2] He did not, as naturalists traditionally have done, simply begin by counting one by one. Instead he employed an insecticidal fog to 'knock down' all the small creatures that lived within the canopies of a single kind of tree, *Luehea seemannii*, in Panama. From this haul, he counted only the beetles; and, in the course of three seasons, he found more than 1,100 species. Then, partly on the basis of common sense, but also employing existing biological knowledge, he began a chain of reasoning. First, he said, it is likely that about 20 per cent of those 1,100 species depend upon *L. seemannii* for survival; about 160 species of canopy beetle. Then, he said, it seems likely (why not?) that each of the 50,000 known species of tropical trees also has 160 species of canopy beetle that are exclusive to it, or at least rely upon it for survival. One hundred and sixty times 50,000 equals 8 million. Thus there could be 8 million different species of canopy beetle in tropical forest worldwide.

God, as I (or rather J. B. S. Haldane) commented earlier, had an inordinate fondness for beetles. In faunas that are known well – such as those of many temperate areas – beetles account for 40 per cent of all arthropod species (where arthropods include all insects, arachnids such as spiders and mites, and crustacea such as woodlice). If the proportion holds in the tropical forest, then the canopy as a whole should contain 20 million different arthropods. The canopy – the leaves and branches – is the main part of the tree: the biggest, the part that is in the sun and air, and which is the most geometrically complex. Even so, the roots and the soils of forest trees which Erwin did not explore should, he reasoned, contain about half as many species as the canopies: a total worldwide of 10 million. Ten million plus 20 million gives 30 million species simply of arthropod in tropical forest. The number for all living creatures should of course be even greater than this. But most creatures are animals and most of them are arthropods; so 30 million represents a good conservative guess for the whole.

Robert May reviewed Erwin's and other, generally more purely theoretical, assessments in *Science* in 1988.[3] He admits that the chain of reasoning that takes us from a study of beetles in one species of tree in one region, to an assessment of biodiversity worldwide, is long. Yet, as he says, there are no obvious flaws in the reasoning. He points out, too, that the figure of 30 million may be truly modest. After all, it seems likely that almost every species harbours at least one parasite that is specific to it; so 30 million immediately becomes 60 million. May points out, too, that beetles are not the only diverse group. They just happen to be the only diverse group that is well studied. Darwin was one of many Victorian devotees: 'Whenever I hear of the capture of rare beetles, I feel like an old war-horse at the sound of a trumpet.'[4] But mites, relatives of spiders which can be smaller than dust (and indeed may be a significant component of dust), may be comparably various. Very few people have looked.

It may seem that conservation biology is fashionable. But if it were one-tenth as fashionable as particle physics or space research, then governments would finance much larger studies, to build on Terry Erwin's preliminary observations. Professor May suggests an albeit teutonic assessment, by teams of biologists, of an entire hectare, with all the species therein.[5] By the standards of conservation biology, this would be a mega-project indeed. By the standards of particle physics it would be almost trivial. Conservation science is urgent, for the species are disappearing. Particle physics is not, for the particles are effectively timeless. There is money in physics, and military technology, just as there is money in Rembrandts. There is not *enough* obvious money in conservation biology, which is why, as was argued in chapter 1, we must bring other values to bear.

Let us take 30 million or thereabouts as a reasonable figure, then (for otherwise we are forced to admit that the true figure may lie anywhere between 5 million and 100 million, a range of ignorance which might truly be considered disgraceful). How many of those species are endangered?

THE EXTENT OF THE THREAT

Most species, as all biologists acknowledge, live in tropical forest. Most individual species (as can be observed from the thousands already known) live only in a limited geographical area; the range of most tropical species is less than that of most temperate species. Neither is this true only of arthropods. Many hundreds and probably thousands of species of fish (so present studies already suggest) each live only in particular streams, which dry up when the tree cover goes (if they are not already overwhelmed by dams). Most of the lemurs in Madagascar are confined to particular pockets of forest, and when the trees are gone, the lemurs go too. Dr Norman Myers has calculated that at least half of the world's tropical forest has been felled since the beginning of this century;[6] and at the present rate of destruction all will be gone by the end of this century. Because most species have only a limited range, a huge proportion of species in any one area are rendered extinct if that area is compromised, for they live nowhere else. Whether there are 5 million species or 100 million, then, it is clear that several millions must already have gone extinct in this century; and that a huge proportion of the rest, perhaps half, which is probably around 15 million, seem doomed to disappear within the next few decades.

You may feel this is all a little too vague, and the chain of inference too long. But again, there are no obvious flaws in it. In any case, we can be more precise. There are two groups of animals on Earth that are well known: the birds and mammals. According to Jared Diamond, a physiologist at the University of California Medical School, Los Angeles but also an outstanding student of tropical birds, we can see that these two groups are in trouble.[7]

The latest edition of the *Red Data Book* of the International Council for Bird Preservation (ICPB) shows that out of the 9,000 species of bird that are known to live on this Earth, 88 have gone extinct since 1600 AD, while another 283 are endangered.[8] Those are large figures; but still they represent only 1 and 3 per cent respectively of the total number, which does not seem too disastrous. But as Dr Daimond told a meeting of the Royal Society in London in 1989,[9] those figures probably underestimate the true position by at least an order of magnitude.

Point one, he said, is that the ICPB – very properly – wants above all to be accurate. So it includes a bird on its *Red Data* extinction list only when all reasonable hope has gone: only after ornithologists have failed to find it after many years' search in the places it is known to have lived. Very occasionally birds that are supposed to be extinct turn up again (Diamond himself found the long-lost yellow-fronted gardener bowerbird in New Guinea in 1981, in mountains that had previously been unexplored). In general, though, the search has to be so thorough and so prolonged that by the time a bird is declared extinct, it very definitely is extinct. More to the point, however, as ICPB recognises, is that by the time the lists are published, they are bound to be out of date.

Recognising this, ICPB's N. J. Collar and P. Andrew in 1988 produced *Birds to Watch* listing the species that might reasonably be thought to be in trouble.[10] They suggest that no fewer than 1,029 bird species are now at risk of global extinction, and another 637 are almost at that point. As Diamond says, when these figures are added to the 88 already known to be extinct, the total comes to 1,754. In other words, almost one-fifth of known bird species of recent times are known to have gone extinct since 1600 AD, or are deemed to be in imminent danger.

Yet even this may be a gross underestimate. The ideal way to compile *Birds to Watch* would be to send out teams of professional ornithologists, worldwide, to count and monitor every species; the same kind of grand sweep that Robert May suggests for exploring an area of tropical forest. But such a search would be impossible with present-day conservation budgets. Again – at every turn - we see that conservation science is under-financed. *Birds to Watch* in practice is compiled from reports of ornithologists who happen to have reason to believe that a particular species is in trouble. Such a system probably works very well in Europe and North America where, as Diamond says, 'there are millions of fanatically devoted bird-watchers' and the rarer the birds become 'the more diligently they are watched'. But most species of bird, like most species in general, live in the tropics, where resident ornithologists are thin on the ground and the birds are often extremely difficult to observe.

Diamond does specialise in watching tropical birds and his own studies 'emphasise the gulf between "proved extinct" and "not proved extinct" '. Thus the *Red Data Book* records that out of 164 native species in the Solomon Islands in the South-west Pacific, one has been rendered extinct since 1600 AD. But Diamond went to the Solomons, looked, read, asked round, and found that there had been no definite records of 12 of the original 164 since 1953, 'even though some of these species were formerly described as common'. Most of the missing birds were ground dwellers and had, it seems, been wiped out by domestic cats. The Solomon Islands are not especially disadvantaged, however. Indeed by the standards of much of the tropics, they are 'a mild example'. Many other areas have far more vulnerable native species, and far more introduced predators.

For all these reasons, ICPB is now introducing 'Green Lists': not to catalogue the most threatened species, with all the attendant delays and uncertainties, but to list those that seem for the present to be safe. But, says Diamond, 'it becomes unlikely that as many as half of the world's recent species will qualify.'

Yet, as I suggested in chapter 1, the present wave of extinction did not start from nothing. We are certainly seeing an acceleration, as technology becomes more vigorous, the human population grows, and people become ever more mobile and widespread. But there is plenty of evidence to suggest that people have been wiping out their fellow creatures for tens of thousands of years.

Evidence from the distant past is mainly archaeological (the things left behind by human beings) and palaeontological (fossils and subfossils), and is difficult to assess. So when we see that some group of animals from the past has suddenly died out, it can be hard to ascertain the cause: whether they were killed by humans, or simply died out because of some shift in climate. Yet, says Diamond, the more that scholars look at the evidence of the past few hundred or thousand years, the more it becomes clear that many animals – especially large animals – have died out, and that humans were involved.

The evidence is mostly indirect, though when it is all put together it is convincing: animals – especially large animals – tended to die out soon after human beings arrived in areas where they had not lived before, and where the animals themselves were 'naive'. Thus, 12 to 13 species of moa died out in New Zealand in the 500 years after the Maoris arrived in 1000 AD. A change of climate might have been responsible – but if it was, it seems a huge coincidence. Besides, there is now plenty of evidence of butchered moas in ancient Maori camps: the remains of 100,000 different birds have now been found. Elephant birds died out in Madagascar soon after the first humans arrived at about the time of Christ. More than half the native birds of Hawaii – 50 species, including a range of flightless geese – became extinct soon after the Polynesians arrived. A similar story is now emerging, as studies progress, in every other Pacific island that has been investigated, from Tonga to Tahiti. Add up all these species, and again it seems that one-fifth of all the world's birds have disappeared in the past 1,000 years: almost certainly, in most cases, because of human beings.

There are plenty of examples among mammals, too. In Madagascar, lemurs once filled all the ecological niches that are now filled in other tropical continents by monkeys and apes. The only lemurs left now are the small tree-dwellers, who resemble monkeys. But once there was one with the size and life-style of a gorilla, and others of various sizes in between.

We find the same story further back in time. The evidence is that the first people reached America from Asia about 11,000 years ago, and migrated down from Alaska eventually to Tierra del Fuego. A wave of extinction seems to have followed them. In North America, 73 per cent of the genera of all large mammals were eliminated.* In South America, the figure was 80 per cent. Again, it is not certain that humans were

*'Genera' is the plural of 'genus'; and a genus is a group of closely related species. The first name in the Latin name denotes the genus: *Felis*, *Canis* or what you will. So each genus that disappears could imply several, or even a lot, of species.

responsible. The climate was changing when people first migrated through the Americas, as the world began to emerge from the latest Ice Age. Again, though, the coincidence seems too great. Again there is direct evidence of human involvement, including the butchered skulls of mammoths. This particular wipe-out has been called 'The Pleistocene Overkill'.

The same pattern is emerging in islands in the Mediterranean. At least ten Mediterranean islands once had dwarf species of elephant, at least some of which survived until recent times, and disappeared soon after human beings arrived. Australia's aborigines, too, seem far from blameless. Most of Australia's genera of large animals – nearly 90 per cent – disappeared soon after the aborigines arrived from Asia about 40,000 years ago. These included a truly giant one-toed kangaroo, a wombat the size of a rhinoceros, and the blade-toothed marsupial 'lion'. Again, humans may not have been directly involved: marsupial lions disappeared 18,000 years ago, at a time of great aridity. Again, though, the weight of coincidence mounts up.

Only in Africa and Eurasia have large animals survived in great variety. These were the continents in which human beings themselves evolved – appearing first in Africa, and spreading to Eurasia from about 2 million years ago. In these continents, the big animals had time to adjust. Even in Europe, though, we have lost such luminaries as the mammoth and the woolly rhinoceros in the past few thousand years. Taken all in all, says Diamond,[11] it seems that human beings, in the past few thousand years, may already have accounted for a half of all the world's large species of mammal. *Inter alia*, there ought at least to be elephants in Europe and the Americas, and rhinos in Europe, as well as in Asia and Africa.

We might choose to speculate that the kinds of forces that brought about all those extinctions of the past are no longer operating. It would be a false comfort. As Diamond says, there have been four main causes of extinction: hunting; the introduction of new species into previously unexposed ecosystems; destruction of habitat; and secondary effects. All four continue.

THE FORCES OF DESTRUCTION

The greatest of the four apocalyptic forces is the destruction of habitats. We may feel that this did not happen in the past to any great extent; but that would be a misreading of history, and of prehistory. The Scottish uplands, for example, are nowadays covered in heather because our neolithic ancestors cut down the oak and pine forest that was there before. The fossils tell us that the Aborigines depleted the megafauna of Australia when they first arrived; and they must also have changed the landscape, with their systematic burnings. (Not necessarily for the worse, but a change is a change.) Easter Island was once forested. The trees were cleared by the people who left the mysterious stone 'men'.

What has changed these days is the rate of change, the extent of penetration. Chainsaws cut trees faster than flint axes; tractors can root out the stumps, where fire merely razed them to the ground. In particular, we are now attacking the tropical forests, which until this century were largely untouched. The forests contain the majority of species; so their destruction is bound to cause the most widespread extinctions. What is worst, though, is that tropical species on the whole tend to be far more localised than temperate ones; so the wildcats survived the destruction of temperate forest in Scotland because they lived elsewhere as well. But to lose a lemur from a pocket of Madagascar is to lose the entire species.

Coupled with habitat loss, and perhaps as pernicious, is habitat fragmentation. Thus the most recent estimates (which are not particularly recent, and are far from intensive) suggest that there might still be between 34,000 and 54,000 elephants left in Asia. But whereas the habitat and the population once ran almost continuously from South-east Asia into the Middle East, it is now divided into scores of tiny patches – only one or two of which, in India, seem capable of individual survival. As the remaining animals in each isolated population grow old and die, the species could yet disappear.

The second most significant cause of extinction, worldwide, is the introduction by human beings of alien species to new habitats. This did happen to some extent in ancient times; thus the Aborigines brought the dingo from Asia, probably about 8,000 years ago, and this exterminated the dog-like 'Thylacine wolf' from mainland Australia. The Romans helped to disseminate the rabbit and the fallow deer. But large-scale import and export of animals and plants did not begin in earnest until the 17th century and beyond. Thus, from California to New Zealand, the native floras have largely been replaced by agricultural weeds brought in inadvertently from Europe, or by European plants brought in for cultivation, and 'escaped'. The domestic cat, which probably jumped ship in Australia in the 17th century, long before Captain Cook arrived, is still laying waste the small marsupials. In much of Australia the cat and the fox, imported in the 19th century for 'sport', have exterminated 90 per cent of the small marsupials. The cane toad in Australia, imported from Hawaii to eat rats in the sugar plantations, is now wiping out many of the native amphibia and small reptiles. Underwater – in rivers and lakes – the effects of alien species may be even more dramatic, though it is largely unnoticed except by specialists. The greatest ecological disaster of this century has occurred in Lake Victoria, where 200 out of 300 native species of haplochromine fish (belonging to the genus *Haplochromis* and related genera) have been wiped out in the past 30 years by Nile Perch, introduced for food.

On islands, introduced predators have proved especially destructive, for island animals tend to be unwary, their populations are small, and they have nowhere to flee. Domestic cats and dogs have wrought havoc worldwide, and mongooses introduced to the Caribbean to control rats have killed much else besides; and rats themselves, which travel wherever ships travel, eat the eggs of nesting birds. Jared Diamond points out that some tropical islands are still miraculously free of rats, but they will appear in the end; and when they do, the extinctions must follow.[12]

Hunting probably accounted for many of the big mammals and birds of the past, as we have seen. Hunting continues, legal and illegal. Only a minority of species is directly threatened; but it is a spectacular list. Some at least of the great whales, notably the blue, may already be so depleted that their decline to extinction may be irreversible. At least 70 of the world's 320 or so species of parrot are now thought to be in danger, and for many of them the trade in 'pets' and feathers is the cause. The Arabian oryx was wiped from the wild by hunters; the last in 1972. Poaching, more than anything else, threatens the future existence of Africa's black rhinos, and adds to the troubles (which of course include habitat destruction) of the four other rhino species in Africa and Asia. Poaching has also reduced the African elephant from several millions in 1980 to about 650,000 today. When populations are small, even small-scale poaching can deliver the *coup de grâce*. There are only 50 Javan rhinos left in the world, and Dr Ulie Seal, chairman of the Captive Breeding Specialist Group of the IUCN, calculates that even if only three of them are poached per year, this would be enough to finish them off. Poaching has to be reduced to zero if they are to have any chance in their native land at all. But where in the world can safety be absolute?

Fourthly, the spread of human beings around the world, with their technology and attendant animals and plants, has all kinds of secondary effects, of which the most conspicuous is pollution. Disasters catch the headlines: oil spills; Chernobyl. More destructive long-term, however, are the smaller effects that may accumulate over decades. The pending greenhouse effect results from pollution of the atmosphere by surplus carbon dioxide and methane; and the effects of heating the world could be as profound as those of the last Ice Age in cooling it. Nitrogen pollution is commonly discussed in the context of run-off, into the ground water; and as oxides of nitrogen from car exhausts contributing to acid rain. But it may be more pernicious than either of these. Research at Rothamsted Experimental Station in England has shown that in industrialised areas, and regions that are intensively farmed, oxides of nitrogen from car exhausts and ammonia from agricultural fields rise into the atmosphere and then descend again upon the land at the rate of 40 kilograms of nitrogen per hectare per year. An arable farmer might apply three times as much or more. But for a forester or gardener, 40 kg per ha is a sizeable input. More to the point is that the wild fields of nature are in general infertile; and most wild plants are adapted to low fertility, and are ousted by weeds (which are the plants that like highly fertile soil) when fertility increases. The effects of these unwonted nutrients from the sky have yet to be assessed. We do know, however, that in Western Australia, one of the principal threats to the patches of remaining bush (with its 9,000 native species of flowering plants – six times the British list) is fertiliser, encroaching from surrounding arable fields.

Finally, we may note that as each species disappears, it is liable to trigger a cascade: others follow that depend upon the one that has gone. The chains of events can take unexpected turns. When jaguars, pumas, and harpy eagles were removed from Barro Colorado island, the ground-nesting birds began to go extinct, because those three 'top predators' preyed upon lesser predators such as monkeys and coati mundis, which in turn preyed upon the birds. Similarly, Michael Soulé suggests that loss of coyotes from the valleys around San Diego has caused native birds to decline, because coyotes helped to keep foxes and cats in check. Such twisting chains of destruction must be occurring especially in the tropical forests, where a huge number of species have evolved in intimate and intricate partnerships with others.

Despite all this, says Diamond, some people dismiss the present wave of extinctions as 'natural'. After all, they argue, all species go extinct sooner or later. I have heard it argued by competent zoologists that black rhinoceroses were 'on the way out' of their own accord; and the same point was made recently in a British national newspaper with respect to the blue whale. Such argument is absurd (as well as distasteful). Neither of those two species would be in any trouble at all if it had not been assiduously shot. Even bad journalists would be rare if there had been a concerted attempt to wipe them out. Besides, as Diamond says, extinctions are now occurring thousands of times more quickly than ever before; many times faster, indeed, than in times of 'mass extinction', such as occurred at the end of the Cretaceous period, and wiped out the dinosaurs. Similarly, Andrew Dobson of London University has calculated that the present rate of extinction is a million times greater than the rate at which new species could appear. Note too, says Diamond, that past species did not necessarily become extinct in the sense that they died out. Many of them simply evolved into new species. *Homo erectus* is extinct. But he and she evolved into us. In any case, Diamond says, 'to dismiss the current extinction wave on the grounds that extinctions are normal events is like ignoring a genocidal massacre on the grounds that every human is bound to die at some time anyway'.

Where though, in all this, is the case for zoos?

IS THERE A CASE FOR ZOOS?

Nothing in this litany of disaster suggests immediately that captive breeding in zoos can play a serious part in the salvation of endangered animals. After all, we have identified the destruction of habitat as the chief cause of destruction worldwide. It seems to follow, does it not, that the protection of habitats must be the antidote?

The arithmetic of the case seems to support that argument. As I will discuss later in this chapter, captive breeding in zoos probably cannot save more than a few thousand species. Yet I have suggested that millions are endangered. Zoos at best, then, it seems, can save only a fraction of a per cent by their breeding efforts. Besides, it is still relatively cheap to buy land in places where wildlife lives; you can buy a lot of Brazilian rainforest for a million dollars. But zoos in general are expensive, and captive breeding programmes in particular can run away with the cash. Money spent on zoos is not being spent on habitats. Captive breeding is not only a waste of money, therefore. It also detracts from more serious endeavour.

It has often been argued, too, that even if zoos managed to perpetuate some species, by keeping them ticking along in an enclosure, that would still be a waste of effort. Animals born and bred in zoos (this argument runs) are fit only for zoos; and if they can never return to the wild they are better off dead. Thus it seems that 'conservation' in zoos, and particularly captive breeding, is merely an exercise in cosmetics; at best, a piece of public relations. Grand expressions like 'conservation breeding' are hollow indeed.

These arguments are to be found in *Beyond the Bars*, edited by Virginia McKenna, Will Travers, and Jonathan Wray, the best modern summary of the views of the anti-zoo lobby known as Zoo Check.[13] The points are easy, beguiling, and widely believed. I believe, however, that they are profoundly misguided, and – because they do harm to an endeavour that has become essential – they are deeply pernicious. But how can such an obvious case be refuted?

In summary, the case for zoos in general and for captive breeding in particular runs as follows. To begin with, nobody, but nobody, doubts that habitat protection is the best strategy for saving animals where and when it can be achieved. However, those who would have us believe that it is the only worthwhile strategy overlook several important points.

First, the whole cause of conservation is chronically starved of money. The safest way to protect a habitat is to buy it (or declare it as a national park, which is effectively the same thing, but employs public money); but conservationists cannot afford to buy all that they would want to. Neither is money the only issue. Conservationists must compete with a hundred other lobbies. In rich countries, in general, conservationists tend to lose out to farmers. In poor countries they lose out in a hundred ways; but in general, Third World governments have tended to feel (and still do) that they have too many other things on their mind to spend too much time on conservation. On land, the pristine wilderness is fast diminishing, and what remains is largely spoken for and earmarked for some human purpose. The best option nowadays, generally, is the national park; but national parks are created by governments, and governments tumble; and people continue to press their claims. Farmers graze their cattle within every national park in India and many of those in Africa (where else can they go?) and the Aborigines of Kakadu in the Northern Territory would like to begin mining. The national parks of Britain are mostly on the uplands (most of the lowlands long since became agricultural) and they are also farmed, forested, and managed for grouse. Even if the best possible efforts were made, then, it still would not be possible to protect more than a fraction of what we would theoretically like to protect; or, in most cases, to devote even the most protected areas to wildlife unstintingly.

Secondly, wild habitats are even more precariously placed than is generally realised. The sheer pervasiveness of environmental pollution is not widely appreciated; for example the fact that wild plants (and the animals that feed upon them) suffer at least as much from excess fertility as from frank toxicity (which is probably much rarer).

War did not feature on my list of four prime causes of extinction, because it is not a prime cause. But we know that the Arabian oryx was severely depleted when the Turks occupied their habitat in World War 1. Of course: soldiers get hungry, and when they are bored they take pot-shots. We know that the scimitar-horned oryx was wiped from its last stronghold in the Sahel by the war in Chad. The last Père David's deer in China were eaten during the Boxer uprising. The Arabian oryx and Père David's deer would certainly be extinct today, and the scimitar-horned oryx very probably would, if they had not been bred in zoos and parks, far from the front lines. In a general way (I know of no formal studies) we know that Ugandan wildlife suffered enormously during the civil wars. The total effects on wildlife of the war in Vietnam is incalculable – but we do know that elephants were deliberately eliminated to deprive the Vietcong of transport. When people are being killed, who cares? But still it would be better if the animals did not die. It is incumbent on those who feel that habitat protection is the only worthwhile strategy to tell us what their plan is for the elimination of war. When they can solve that, then we can put our faith entirely in national parks.

There are no absolutes, either, in this untidy world. We may create a national park; protect it with a dozen treaties; monitor and exclude the pollutants as best we can; replant the missing trees; and yet for a dozen reasons of biology, which have to do with area, and populations, and genetics, we still may find (and on continents generally will find) that the species continue to go extinct.

Certainly, if we turn an island into a national park, in its entirety, then it may, with luck, continue much as it did before. True, the seas around it may not be quite so rich in fish as once they were. The whales may not come as frequently, or the turtles in such numbers, as once they did. An oil-slick may happen by. But if we *could* give an island perfect protection, it should continue to flourish.

But if we create a national park within a continent, then we are, in effect, creating a new island; one that is surrounded by cities and farms rather than by water, but an island none the less. Islands are small havens in the midst of hostile territory, and as such, they confer some advantages on their inhabitants, but they also pose many special problems. The animals that habitually live on islands have evolved to live on islands, and have solved those problems. Continental animals that suddenly find themselves in reserves that are effectively islands are not so adapted.

There are three kinds of problem. First, if you look at any animal's habitat through the eyes of the animal, you find that it is not, as it may appear, a *tabula rasa*. It is, rather, a mosaic: a series of particular and necessary niches; a place to feed, a place to rest, a place to find mates. It is not always obvious to the observer what all the components are. We can be reasonably sure, however, that if an animal lives habitually on an island, and never leaves that island, then that island must – logically – contain all the niches that it needs.

But we cannot have any such certainty when we carve out a fragment of a continent and call it a national park. What of the animals that normally migrate to pastures elsewhere? Sometimes this is obvious – as with the antelope and zebra that habitually seek fresh pasture outside the national parks of Africa. With small birds and the myriad invertebrates, we simply do not know what they once sought outside. We can only watch them disappear, despite our best efforts.

Then again, despite our best efforts, we cannot in reality prevent the degradation of these continental 'islands'; we cannot truly prevent the encroachment of the surroundings. Fertiliser and its attendant weeds are among the most obvious invaders. We can of course take steps to make good the components that are missing; water is laid on in Kruger

National Park in South Africa, for example, and this (plus electricity and the rest) is commonplace in parks these days. But often we do not know what is missing until too late, or at least until destruction is well advanced. Thus, there are still plenty of trees in Western Australia, and cockatoos nest in trees, so everything seems fine. But they nest in hollow trees; and foresters remove hollow trees, in the interests of creating 'healthy' forest. There are still plenty of cockatoos, but that is because they live a very long time. Many of the remaining birds have not bred for some years, and unless they are equipped with nesting boxes very soon, they will disappear. Pandas in China also suffer from a lack of hollow trees to rear their young. By the time the present generation of trees have had time to rot (assuming they are allowed to do so), the animals will be extinct. If boxes are introduced for them (and water) then we may legitimately ask (as I will ask again in chapter 8), 'Where does habitat protection end, and captive breeding begin?'

Population is perhaps an even bigger problem. As will be discussed again in chaper 4, populations need to contain several hundred individuals at least, if they are to be viable in the long term. Less than that, and they will die out from accident or from inbreeding. But a large population of big mammals requires an enormous amount of room; obviously, because each individual eats a great deal. On islands – real islands – you simply do not find large mammals, except for seals and sealions, which feed at sea. The ancient elephants of the Mediterranean islands evolved to be tiny, no bigger than sheep; and besides, the islands that harboured them, such as Malta and Cyprus, were not especially small. Almost all the surviving big animals on small islands are reptiles: giant tortoises; Galapagos iguanas; Komodo 'dragons'. Reptiles, weight for weight, eat only about a tenth as much as mammals.

Yet nowadays we expect animals such as elephants to cram into national parks and into even smaller reserves. Elephants in particular not only eat a great deal, but are destructive to boot. They are gourmets, who will fell a tree for a succulent bud. When they had Africa and all of South Asia to themselves, there was no problem. Perhaps their numbers rose and crashed over centuries, as those of lemmings do over a few years. Perhaps they laid an area waste and then moved on, like swidden farmers. Biologists hold various opinions. What is clear, however, is that boom-and-bust swings in population, or swidden farming, cannot be sustained in parks that are smaller than continents. Elephants in parks must be rigorously 'managed'.

Elephants, too, like many big animals, breed only slowly. Their populations boom in time, but only after decades or centuries. If their numbers are depleted, it may be years before they reach 'safe' levels again. For every month that passes, the danger of total obliteration is extended.

Of course, elephants are especially big, and particularly slow-breeding. With other animals, we may suppose, the problems are easier. But elephants do have one advantage: they are herbivores. Large carnivores need even more space, because they need entire herds of large herbivores to feed upon. How much space is needed to maintain a viable population – many hundreds – of tigers? More than is available in Sumatra, that's for sure. Perhaps there is enough room in India, provided by the reserves in Mrs Gandhi's Project Tiger. We shall have to wait and see. But if there is, then India will be the only country with viable populations of wild tigers. The Javan tiger disappeared in the 1970s. The Balinese has gone; the Caspian and South Chinese have probably gone. Wild dogs need huge areas too; and some biologists suggest that the only viable population left in all of Africa is the one in Botswana. Yellowstone, jewel among North America's wonderful national parks, may yet prove too small for its grizzly bears.

A third problem, a variation of the theme of population, will be discussed in more detail in chapter 4. In essence, though, it is that many wild populations are in reality divided into smaller sub-populations, which do not have much to do with each other. Unless detailed preliminary studies are carried out; and unless conservationists are given the freedom to buy precisely the piece of land that they want; then it is quite likely that a reserve will be created across the borders of two different sub-populations, instead of in the territory of either one. In such a case, both sub-populations may be too small to survive, and both will go extinct. In general, because field biolgists are thin on the ground, they do not analyse in detail precisely the reasons why animals invariably go extinct even after they are apparently protected within reserves. All that is certain is that the extinctions do continue, and may continue to do so for several decades. This period of continued extinction is known, somewhat quaintly, as 'species relaxation'.

With all these problems in mind, ecologists are now discussing the ideal shape of wildlife reserves. If the national park is one big patch, then that reduces edge effects; but it may mean that necessary components of habitat are excluded; and that much land is included that is less than ideal. If the park is divided into small areas, each on land that is good for wildlife, then the population in each patch may be too small to be viable. Perhaps the ideal is to create a network of patches, linked by 'corridors' of habitat through which the animals can migrate. It is not known, however, whether animals would use such corridors. There are no first principles from which to predict this. Some birds are happy to fly over open ground, and do not need corridors. But some are not. Besides, some biologists, such as Dr Dan Simberloff at the University of Florida, argue that corridors could well be dangerous. Islands (comparable with small patches) are not all bad, he says. The animals on them are protected from predators and diseases that come in from elsewhere. Corridors could expose animals to threats they would not otherwise be exposed to. The point is that debates such as this matter, because conservationists can afford to buy only a limited area of land; and they need to ensure they deploy their limited wealth in the best possible way.

To be sure, the pending greenhouse effect gives this discussion a new edge. We cannot be sure precisely what the effects will be, or how extensive. But if things work out the way the scientists are suggesting (and if they do not, then there is something very wrong with some fairly elementary physics) then we can be fairly sure that the climate of any one place is liable to change. Most places will become hotter – though some could become cooler, because, for example, ocean currents could change direction. Some places will become drier, and some wetter. If a national park that is now established in a wetlands area loses its water supply, what then? If a temperate forest becomes subtropical, what happens to the animals that are adapted to the specialist trees?

Of course, there have been dramatic changes before, for example during the Ice Ages; and also over much greater periods of time, as the continental land masses drifted through different latitudes. But during those changes, we know that animals had a tendency to go extinct. We also know, however, that animals of the past were favoured by three factors that are not on the side of present-day animals in reserves. First, their populations were in general much larger than today; tens of millions of individuals in some cases, as opposed to a few hundred or thousands today. Second, the big changes happened relatively slowly: generally over centuries, or even millenniums. Third, as conditions changed, animals (and tree-lines) migrated. Many were saved from extinction just by moving their feet. Eighteen thousand years or so ago there were reindeer in the South of France, and now they have gone back north.

The changes wrought by greenhouse could in theory be just as dramatic (though in reverse) as those of the Ice Ages. But they will come on quickly; far too quickly for animals to evolve a genetic response. And if the animals are trapped in reserves, where will they walk to?

From all this discussion, four points emerge. First, although habitat protection is difficult – far more difficult and precarious than most people realise – we cannot give up on it. It must remain the priority. Every opportunity to protect wild places must be taken. Even with the best will in the world, however, we cannot save all animals by habitat protection. Some habitats cannot be protected at all; and those that can be are liable to be compromised in ways that we can in practice do nothing about, for we can do nothing about other people's wars, or the realities of population dynamics. Captive breeding can save some of the animals that cannot be saved in the wild.

Thirdly, an increasing number of wild populations that are apparently thriving in reserves will be 'on the edge'; their numbers just above what is needed to be viable, but liable to dip dangerously low if they are hit by forest fire, or epidemic, or (like the last remaining population of native Puerto Rican parrots) by a hurricane. Captive populations can keep animals in reserve, to boost the indigenous populations when they flag, and help them over the hump. I will discuss this possibility again in chapters 5 and 8.

Finally, note that whereas millions of Amazonian beetle are becoming extinct because their habitat is being destroyed, most of them could be saved if their habitat was reasonably protected. For a growing catalogue of large animals, such as the Sumatran tiger and possibly for the Sumatran rhinoceros, this is simply not true. To be sure, if we banished the human population of Sumatra, or at least curbed the growth of population and replanted the trees post-haste, then they could be saved *in situ*. But these things are not going to happen; and indeed, people are now being 'transmigrated' *into* Sumatra. In other words, for more and more big animals, habitat protection would not be enough even if it was carried out as well as is feasible. For the Sumatran tiger, habitat protection alone probably cannot succeed; and though Sumatran rhinos seem safer, they too may also need captive breeding.

What of the arithmetical objection – that captive breeding can save only a few hundred or a few thousand species, while millions are endangered? I will discuss later in this chapter precisely how many species might be saved. But a few preliminary observations are in order. First, the arithmetic is of course correct. Millions of species are in danger. Captive breeding can save only a few hundred, or at best a few thousand. But does that really mean that it is not worthwhile? Consider just a few of the issues. First we may note that the animals for whom the very best habitat protection that is feasible may not be sufficient are, in general, the large land vertebrates. In short, the animals for which captive breeding is most obviously justified are the rhinos, the tigers, leopards, primates, parrots, probably the Asian elephants, many an antelope, many a bird of prey, various cranes – and so on; all the creatures of our childhood; what most people mean by the word 'animal'. These are the creatures captive breeding is equipped to save. Are the critics really suggesting that these should be written off?

The point is made; yet we can pursue it further if only in the interests of logic. The critics suggest (in effect; whether they realise it or not) that we should not attempt to save the Sumatran rhinoceros by captive breeding because we cannot in practice save every Amazonian beetle by captive breeding. Is that a sound argument? Its deep lack of logic becomes evident if you state the argument the other way round. Thus we could say, 'With the best will in the world, we cannot save the Sumatran rhinoceros simply by habitat protection. Therefore we should not save Amazonian beetles by habitat protection.' Of course (you never know when you are going to be misquoted in this business), nobody *would* say that.

But in any case, why make a conflict of this? Habitat protection is vital, of course. It is indeed the chief of the conservation strategies. Nobody denies this. Captive breeding is of lesser significance, but it still has an important role to play; not as an alternative to habitat protection, but increasingly as an adjunct to it. Habitat protection is all that some species require. It is not sufficient (as things are at present in the world) for all species. Captive breeding can save many of the species that cannot be saved by habitat protection alone. So – horses for courses.

Still, though, the critics ask, why zoos? And why keep animals from Africa or Asia, in Europe and North America? After all (as Virginia McKenna points out [14]) there are reserves in Africa that specialise in raising and protecting black rhinoceros. I have fond memories of De Wildt, near Pretoria, which specialises in cheetahs. If we are to proceed with captive breeding, then surely it is best to keep the 'captive' animals in their own kind of country, and quasi-wild environments?

Again, though, there is no need to turn such an argument into conflict. Reserves especially created in the endangered animal's own country have many obvious attractions. On the other hand, it can pay, too, to keep several species together, not least for reasons of economy. This can make best use of space, or wardens and keepers, of vets and scientists and secretaries. Of course there are advantages in keeping animals in their own country, and there seems, for example, to be very little case for taking animals such as koalas out of Australia. But sometimes the same threats that drove the animal to extinction in the wild continue to harass it in captivity. Security against poachers remains a problem in Africa's rhino reserves. It is less of a problem in Europe and the USA. A reserve in Chad would not have saved the scimitar-horned oryx. Hurricanes are a feature of Caribbean islands; and Hurricane Hugo in 1989 would not have wiped out half the remaining Puerto Rican parrots, if they had been somewhere else. We might argue in a general way that tropical animals ought at least to be kept in the tropics. But does Africa want to be burdened with the rhinoceroses of Asia, when it has trouble protecting its own? If Africa was rich and had time to spare, and if Europe was poor and at war, then it would be good if Africa kept a few European wildcats and sand lizards, to tide them over. But the world is not like that at present. For the time being, it makes sense for Europe and North America to keep some African animals, while the same arrangement the other way around would make much less sense. That is just the way things are. But if you put several species together and curtail their freedom, and perhaps take them to a foreign land, what do you have but a zoo?

What of the argument that captive-bred animals can survive only in captivity? That simply is not true. As a generalisation it is obvious nonsense; after all, every species that human beings have ever taken into domesticity has at various times returned successfully to the wild. The list includes cats, dogs, goats, sheep, cattle, horses, mink, budgerigars, ostriches – and so on and so on. But domestic animals are bred and trained not to be wild; while captive wild animals, these days, are encouraged not to be domestic. It is true that some wild species find it harder to return than others; but that is hardly suprising, because some are brighter and more versatile than others, and some live in more difficult habitats than others. In general, there seems no reason to suppose that there is any captive-bred animal that could not be successfully returned to the wild, if it was properly prepared. At present, about 100 schemes are in train worldwide to reintroduce captive-bred animals to the wild. I will discuss this at greater length in chapters 5, 7, and 8. In general, every zoo conservationist envisages that return to the wild is the natural and the proper end-stage of their endeavour.

Finally, a general case for zoos that was made long before the need for captive breeding was appreciated, is that they serve as sources of knowledge. To be sure, such an argument does not justify the zoo unless that knowledge is used for conservation, rather than for agriculture or medicine. But the knowledge certainly can be used to aid conservation. For example, it may never be possible in practice to keep Asian elephants in Europe or America in sufficient numbers to save the species. But if Asian elephants are to be saved at all, then they must at least breed in captivity in Asia – because an increasing proportion of Asia's elephants are captive, and are kept for work. Yet they do not breed well in captivity in Asia, any more than they do in the West. So one essential task for zoos is to find out why they breed so erratically, and to overcome the problems. It is difficult to see how the necessary research could be carried out in a logging camp.

This, then, is the general case for zoos. Nobody suggests that captive breeding is a substitute for habitat protection; but for many species it is at least a necessary adjunct, and for some may offer the only immediate option. Zoos are not the only places in which animals can be bred in captivity, but they have advantages, as well as disadvantages. Zoos do not necessarily turn wild animals into domesticates. They can, and increasingly do, prepare them for a new life in the wild, once the wilderness has been made hospitable to them again. They cannot save every endangered species, but the ones they can save are worth saving.

We have been vague, though, on the issue of numbers. How many endangered species is it possible to save by captive breeding? And – to put the matter another way – what proportion of the species for which captive breeding is a necessary strategy can in practice be saved by it?

WHAT CAN CAPTIVE BREEDING ACHIEVE?

In practice, as I have argued, captive breeding can make its greatest impact among land vertebrates: mammals, birds, reptiles, and amphibians. Among the committed aquatic mammals, some sirenians (dugongs and manatees) are beginning to make a showing in captivity, but the cetaceans (whales and dolphins) are not and it is hard to see that captive breeding can contribute a great deal to their future. There is certainly a case for saving many a fish in captivity, on an *ad hoc* basis; and for breeding at least a select list of invertebrates. I will discuss this below, and again in chapter 5. In general, though, the list of endangered species among fish and invertebrates is so great that captive breeding can probably save only a small proportion; but a small proportion is far, far better than none at all. Land vertebrates, however, are the principal target.

And one of the main theoretical objections to captive breeding – that it can save only a minuscule proportion of endangered species – falls away, once we home in upon the land vertebrates. For vertebrates as a group are nothing like so diverse as invertebrates. Biologists agree that there are roughly 4,000 species of living mammals; 9,000 species of bird; 5,000 reptiles; 2,000 amphibia; and somewhere between 25,000 and 40,000 species of the various kinds of creature (sharks, rays, lung-fish, bony-fish) that are called 'fish'.

At the Workshop on Genetic Management of Captive Populations in Virginia in 1984, Michael Soulé and colleagues suggested that the total of mammals that is likely to require maintenance in captivity in the next 200 years is just over 800; including 100 or so out of the 900 species of bat; all 160 species of living primate – the apes, monkeys, and lemurs and their relatives; all the 35 living dogs, including the wolves and foxes; 60 out of 72 cats; about 100 of the 172 even-toed ungulates, which include the cattle, antelopes, deer and giraffes; all 15 species of odd-toed ungulate (rhinoceroses, horses, and tapirs); both the two living elephants; all four of the sirenians, the manatees and dugongs.[15] The figure of 800 becomes much larger if we break the species down into subspecies; for example, there are eight recognised subspecies of tiger (one of which is the Siberian) though in the above list this appears as only one. Clearly, then, sharp decisions still have to be taken as to which subspecies should be saved, or whether indeed the different subspecies of the same species should be merged; a subject to which I shall return. Even so, though, 800 does seem a manageable number, for all the world's 1,000 or so zoos to manage between them.

The same kinds of considerations apply to birds. Probably, as Jared Diamond argued,[16] a half of them are in trouble. For many at least of those, habitat protection if vigorously pursued should be sufficient, and is probably the best option. This would still leave a few hundred that would require captive breeding if they are to survive. The same kind of arguments apply to the other land vertebrates, the reptiles and the amphibia. In general, then, Soulé and his colleagues agreed, as a rough working figure, that about 2,000 species of land vertebrate would need captive breeding in the next 200 years. (The significance of this '200 years' will become apparent later.)

For zoo populations truly to contribute to conservation, they must be 'viable'; which essentially (when the word is applied to populations) means self-sustaining. Ideally, indeed, they should provide a surplus, to feed back into the wild. For this, as I have hinted, each population generally needs to contain several hundred individuals: not all in one place, but spread around among different zoos. Dr William Conway, director of the Bronx Zoo in New York, calculates that if all the zoos in the world devoted half their present space to breeding endangered vertebrates, then they could, between them, support about 800 viable populations.

Eight hundred is not 2,000. But it is of the right order of magnitude. It does not fall absurdly short of what is theoretically needed. Dr Michael Brambell, director of Chester Zoo in England, points out indeed that if all the world's zoos truly put their weight behind captive breeding, and if the public put their weight behind the zoos, then they could between them save all vertebrate species that are likely to need captive breeding in the foreseeable future. That would be a very significant achievement.

There remain some vertebrates that cannot reasonably be saved by these means. The cetaceans – whales and porpoises – certainly do not breed well in dolphinaria; and any reserve that was earmarked for them would have to be so extensive that they would hardly be captive at all. But the fears that have surrounded the fate of many land vertebrates in captivity look less and less valid. As we will see in later chapters, even elephants and polar bears, favourite targets of anti-zoo lobbyists, can be kept humanely and breed well, if enough money and ingenuity are extended.

For fish and invertebrates, captive breeders will just have to do the best they can. Freshwater fish in particular are endangered. Like forest beetles, many species tend to live only in particular places: isolated ponds or streams. The threats are horrendous; drainage, dams, pollution (for ponds are sumps of pollutants in all the catchment area) plus sheer carelessness, for anglers, fish farmers, and governments alike are apt to introduce exotic species with no thought for the native inhabitants. I have already mentioned the many scores of species of haplochromine fish eliminated from Lake Victoria by introduced Nile perch, and will discuss attempts to breed them in captivity later. In practice, captive fish and invertebrates do have one advantage over the breed and antelope. True, they cannot hope to make numericall' inroads on the entire endangered faunas. But they can more ea٭١، maintain large and viable populations of each species. Small aquaria and vivaria could in theory save entire species.

Yet there is one last general objection to conservation by captive breeding. Suppose we do manage to maintain some species of monkey or parrot in captivity after its habitat has gone. What is the point? An animal with no home to return to is a sad creature indeed. There are two strands to this argument: first, that captive-bred animals simply are unable to return to the wild; and secondly, that even if they could, there is nowhere for them to return to. If there were (the argument goes) they would not have become extinct in the wild in the first place. As will be discussed in detail later, the first of these arguments simply is not true. Some captive-bred animals return more easily to the wild than others. But some find the transition ridiculously easy (including domestic cats and dogs, which have many times successfully become feral); and there is no reason to give up on animals such as the orang utan, which clearly find the return more difficult.

But we should discuss the second point – that there is nowhere for captive-bred animals to return to. The response to this point is a crucial component of modern conservation philosophy.

THE FINAL RETURN

Animals are being driven to extinction through several causes, but one fact underlies all of them. Human numbers are growing. It is not just a matter of numbers. Consumption matters too. As Dr Paul Ehrlich of Stanford University is fond of saying, well-regulated families of two adults and two children in Los Angeles or Berlin may feel that they are environmentally unimpeachable; but they probably consume more as a family than an entire Bangladeshi village. Mobility matters, too. The severely threatened creatures of the Mediterranean (including the small, shy monk seal) would not be in so much trouble were it not for tourists. But numbers are still the overwhelming threat.

The rise in numbers, or at least the rate of rise, must decline soon. At present, world population is doubling roughly every 30 years. It will almost certainly reach 6 billion by 2000 AD. It would be 12 billion by 2030 AD, if the rate continues: 24 billion by 2060 AD; 50 billion by the end of the century. Optimists here are those who feel the more the merrier, and that technology will find a way. But much of the world, including the most fertile parts, are already crowded. I have spent a great deal of time talking to agriculturalists over the past 20 years and have never met one who thinks that 50 billion would be sustainable; or even 20 billion, for more than a few years. Something has got to give.

We cannot dismiss the possibility of a horrendous crash. But there is an optimistic demographic scenario. In conservation, as Michael Soulé says, it is *necessary* to be optimistic. Despair is the only alternative, and that would bring the end of all action. Hope, said St Paul, is one of the three necessary Christian virtues; and it is a comment of enormous subtlety. The optimistic scenario is that people will begin to curb their own population growth by benign and voluntary means. After all, when people become rich, and societies become more egalitarian (so that people do not feel they have to have children simply as a way of achieving status), they sometimes (though not always!) choose to have fewer children. If they have only two per family, then the population must eventually stabilise and indeed come down; for some of the offspring will be infertile, and some will choose to have no offspring of their own, and some will die before they have the opportunity to breed.

It is possible, then, that by benign and voluntary means the human population will stabilise at 8 to 12 billion in the middle of the 21st century. If it does, the human species may pull through without a crash (or at least without a crash on the scale of billions). Furthermore, if human beings as a whole eventually come to realise that unlimited breeding is not an option; and if indeed they come collectively to accept that life is better and safer if there is elbow room than if there is not; then, continuing the two-children-per-family policy, the population would begin to come down. If it does not, then human beings will be in huge trouble, because there is good reason to doubt whether 8 to 12 billion is sustainable.

But if the numbers do begin to reduce, then people could begin to ask what number would be ideal? At what point should we halt the decline? Michael Soulé has suggested that the optimum number would be about 100 million – the world population at the time of Christ which was, as he says, 'a time of towering genius'. Paul Ehrlich feels that 1 to 2 billion would be a comfortable number. Note that both of them are humanitarian. They are not saying they would like to see fewer people because they do not like people. Their point is the opposite: that fewer people could lead lives that were more fulfilled, with more options; that enormous populations seem bound to crash; and that if fewer people live at any one time, then eventually there will have been more people on this Earth than would have lived if they had all been around at the same time, because if the latter were the case, then the entire species could come to a precipitate end.

Future generations can decide for themselves, however, what the final number should be. All that is certain is that numbers must reduce, and the optimistic view is that they will reduce because people choose to reduce them, and not because they are overtaken by disaster. Even the optimistic scenario will take several centuries to unfold, however. For birthrate is not the only determinant of population growth. Death rate matters too; and at present death rate (per thousand population) is going down because people are tending to live longer. Because of this counter-force, it would be 500 years – and probably nearer 1,000 – before the world population declined again even to present levels.

What does all this imply for the conservation of animals? At present, we are making life intolerable for wild animals, and in particular for large ones. That state of affairs is liable to persist at least for 500 to 1,000 years, a time that Soulé has called 'the demographic winter'. After that, however, life could in theory become progressively easier, as human numbers continue steadily to decline.

Even now (as will be discussed later) it is becoming possible to return at least some animals to the wild, as isolated pockets of wilderness (or rather, of managed reserve) become available. In 500 to 1,000 years, if all goes well, wholesale return might be possible. The overall plan for captive breeding, then, is to maintain as many species as possible for 500 to 1,000 years. That is a long time, but it is only a twinkling compared to the total lifetime of most species in a state of nature, and less than a twinkling compared to the total life left to this planet. The species that could theoretically be carried through the demographic winter by captive breeding include all the land vertebrates.

In practice, as we will see in the next chapter, most captive breeding plans are scheduled for technical reasons to last 200 years, rather than 500 to 1,000. The general point is, however, that the critics of captive breeding are misinformed. Life in captivity is not intended to be the end of the line. It is an interim state: prolonged, but interim none the less.

You may feel again that the chain of reasoning in this chapter is too long: that the grand demographic projections are too sweeping; that '500 to 1,000 years' is too great a swathe of time to think about sensibly. The truth is, however, that if we want conservation to be more than a pastime for the under-employed, it has to be planned; we have to get used to thinking in long periods of time, and to speculating through long chains of reasoning. The alternative is to cross our fingers, and hope for the best. That to be sure is the attitude of most governments and most politicians. But it cannot be the best we can do.

In truth, it may not prove possible to save more than a small proportion even of the vertebrates that are now endangered. If we save just one, however – the Javan rhino, the Persian leopard, the Puerto Rican parrot – it will be better than saving none at all. If the Louvre was on fire, it would be better to pull a few Leonardos from the flames than to allow them all to burn. If we make no effort, or no properly planned effort, then we can be certain that most or all of them will disappear.

The argument that captive breeding detracts funds from other conservation strategies is simply unfounded. In general, societies spend ridiculously little on conservation, and the task now is not to deplete captive breeding so as to augment habitat protection, but to seek to increase expenditure on all sensible conservation measures by ten- or 100-fold. Captive breeding centres – zoos – can help in this too, by focusing attention on the problems.

This, then, is the case for captive breeding. The first essential is to ensure that the captive animals are happy, healthy, and breeding well. That is the subject of the next chapter.

THREE

FIRST GET YOUR ANIMALS TO BREED

A hundred years ago – or even a decade ago in many cases – the life of animals in zoos could best be described in the words of Thomas Hobbes: 'solitary, poor, nasty, brutish, and short'. Now, curators of good zoos can effectively guarantee – barring hurricanes and other Acts of God – to keep animals alive in captivity in most cases for far longer (perhaps several times as long) as they could reasonably expect to live in the wild.

Nutrition and feeding have largely been sorted out (even though the findings are not yet universally acted upon). On the broad scale, for instance, curators now appreciate the difference between ruminant animals such as cattle that are rough grazers and can thrive even on a diet of straw, and fastidious browsers such as giraffes which naturally feed on the tenderest protein-rich treetop leaves, and whose rumens are damaged by coarser fare[1]. They appreciate many quirks of detail, too. There is the high requirement of animals such as Grevy's zebra and black rhinoceros for vitamin E (a compound that protects tissues against excessive oxygen, and is vital for animals that have a 'fight or flight' mentality and use lots of oxygen in short bursts, and whose natural diet is rich in E). There is the need of all animals for vitamin D, which many land animals synthesise when exposed to sunlight – so that animals from monkeys to tortoises may be given a daily dose of sun-ray lamp. There is the appreciation that cats must eat animal flesh, because they have a special need for particular unsaturated fatty acids that only flesh contains; although carnivores of all kinds (cats, vultures, what you will) suffer if they are given only red meat, for they also need offals for the vitamins and chips of bone for the calcium[2]. The nutritional science of exotic animals continues to expand, but most of the problems now seem soluble.

Infections are a perpetual threat. The Romantic conceit that wild animals are naturally healthy is simply not true – infections from phocine distemper of seals to rabies of all warm-blooded species, from avian tuberculosis to the universal roundworms, plus a host of ectoparasites and poisoning (for example from botulism) – play an enormous part in the lives of wild animals, and indeed play a significant role in regulating wild populations. The risk of infection is to some extent reduced in zoos (because, for example, an animal might escape the vectors that carry the disease in its own country – as monkeys in northern Europe might be spared mosquito-borne malaria). But other dangers are increased, as animals may be more closely packed than in the wild, and seem in general to be more likely to catch disease from other species or from human beings. Malignant catarrhal fever, for example, may spread in zoos from sheep to deer (in which it is liable to be lethal); and captive apes may catch human diseases from colds to measles. These days, however, the veterinary care of animals in the best zoos is excellent, and the animals within them are freer of parasites than any wild animal has ever been.

Because vets now handle disease so confidently, curators can feel more relaxed, and architects can design with much more freedom. Thus, until a few decades ago zoos felt obliged to provide cages with scrubbable floors and tiled walls that could be washed down with disinfectant. The Joseph Lister approach to husbandry still persists in some quarters, but curators for the most part are happy these days to provide earth and trees and woodchips: not sterile, but far friendlier. The only modern trend that I personally do not like, is for keeping animals in painted boxes, with fibreglass trees and painted backdrops, that *look* very attractive to the visitor but may be barren from the animal's point of view. A painted backdrop is probably better than a blank wall; a fibre-glass tree is better than a bare floor; and an animal that is asleep anyway might as well rest against a painted sky, provided it also has the chance to do more interesting things at other times, and real vegetation rarely lasts long when exposed to the concentrated attentions of animals. So fibreglass and backdrops do have some role to play, but they are not a substitute for the real thing. They lack the smells, textures, variety, and unpredictability of nature. At London Zoo, for example, various small monkeys have recently been given lawn-sprinklers to simulate showers – which most of them love. After the shower, they lick the water from the natural leaves, but not from the leaves of plastic plants that are included to thicken the foliage. They can tell the difference.

In general, curators these days try more and more to simulate wild conditions in captivity. This is good for the animals' welfare – for what better criterion of well-being can there be, than the animals' freedom to live their natural lives? It also means that animals in captivity should retain the capacity to return to the wild. Reintroduction has become an essential component of conservation, as discussed in Chapter 5; and the methods now employed specifically to conserve the natural behaviour of captive animals, are discussed in Chapter 7.

With good feeding, freedom from disease, and friendly accommodation, there might not seem much to prevent animals from breeding; which, from a conservational point of view, is what is required. Two points arise, however. The first is that there is no simple relationship between reproduction and welfare. We cannot take it for granted that because an animal is breeding well, it is therefore happy. Human beings breed under the most dreadful circumstances. Farm animals breed like no animal has ever bred before, though many are kept under conditions of extreme behavioural deprivation. In zoos, roan antelope have not done well up to now, and mortality has been high – though Whipsnade, in England's Bedfordshire, seems to have solved the problem by giving them plenty of opportunity to keep out of the public gaze. Yet roan antelope do breed reasonably well in zoos.

The second problem, however, is that animals do not take lightly to sexual reproduction, for it is innately hazardous. Males and females alike chance their lives (and sometimes lose them!) by risking such close contact with each other. They also run enormous risks in finding mates, for the process is at best time-consuming (and time spent courting is not spent feeding); and some animals have to fight for mates (which is extremely risky) while others display to attract sexual partners and thus, also, attract predators. Animals *in flagrante* are also especially vulnerable.

For all these reasons, all species of animals have evolved elaborate codes and rituals to ensure (a), that their intended sexual partner is indeed of the appropriate species (and indeed, probably, of the appropriate genetic make-up within that species); (b), that the intended partner is of the appropriate sex (which in the case for example of some penguins is not always obvious even to penguins); and (c), that the intended partner is sexually receptive (for if not, then sexual advance is common assault, and the price of common assault may be death). If these conditions are not met, then no mating takes place. If wires are crossed, there may be a very serious fight instead.

Thus, in addition to comfortable quarters, curators who intend to breed animals must provide conditions that allow the rituals and codes to be enacted; and these conditions are just as much social as physical. For example, golden lion tamarins and common marmosets are among the group-living animals (wolves and naked mole rats are others) in which only the dominant male and female breed at any one time. The non-dominant adults and sub-adults are 'sexually quiescent'. Thus breeding of these animals is very slow unless pairs are isolated. Among woolly monkeys, on the other hand, the dominant males bully the females unmercifully unless they have an entire troop to look after, with several females and juveniles; so group-living is the key to success.

Many animals will not breed if they are too familiar with their intended partner – if, indeed, they have reason to 'suspect' that their intended partner is a sibling. Cheetahs did not breed in captivity until (among other things) it was realised that males and females should not be brought into contact until the female is in heat.

Among many species, a little male rivalry is required to get them into sexual trim, and indeed sometimes to excite the females. One reason for the many disappointments in breeding elephants in captivity, so many believe, is that few zoos can afford to keep more than one bull; and bulls may not mate (nor females find them interesting) unless there are a few other males to jostle, and (for the females) a choice. Breeding of white rhinos may have failed very often in the past because although they live in herds, with males and females constantly together, the females undoubtedly prefer to mate with new males, who they are not in contact with every day. They are very responsive to the smell of other rhinos, as carried in urine and faeces; and trials are in progress (in a cooperative venture between the Institute of Zoology in London and Dvur Kralove in Czechoslovakia) to see whether female white rhinos can be made sexually receptive simply by bringing the excreta of a novel male into their paddock. Giant pandas, too, may need a little male rivalry to get them going. Certainly, captive breeding has not been easy so far; but females in the wild attract a whole bevy of suitors, who spat and slug it out before one of them mates. Intelligent animals may also be very choosy at a personal level. If two animals that are brought together for mating simply do not like each other – well; too bad. When animals are kept for practical reasons only in small numbers (such as elephants) such choosiness can hold up the breeding programme entirely.

Some birds, in particular, breed all together, in a rush. This is generally thought to be a survival tactic; predators take a smaller proportion of the young if thousands are born all at once, than they would if breeding was staggered through the summer. This implies, however, that these gregarious birds also mate at the same time; and some will not mate at all unless they are surrounded by many others of their own kind. But it is difficult for zoos to keep animals in huge flocks. The Wildfowl and Wetlands Trust at Slimbridge, Gloucestershire, fitted their flamingo pen with mirrors to make the birds think they were in big company.

There are still breeding problems to be solved. Elephants still do not breed reliably in zoos, and neither do rhinoceroses, although all of the black rhinos of breeding age in Britain are either pregnant at the time of writing or have given birth in the past year or so. Giant pandas are still a great challenge. Clouded leopards are a nightmare. They have huge canine teeth – by far the biggest, for their size, of modern cats. The males are much bigger than the females. When put together to mate, the males tend simply to attack. What happens in the wild, nobody knows; but many a captive female clouded leopard has been killed during an intended mating. One zoo in my ken keeps from the public gaze a three-legged female clouded leopard who lost her leg in a mating accident. Some curators feel that artificial insemination is the only reasonable way forward, but others feel that the husbandry ought to be good enough to solve the behavioural problems. Whatever the solution, the fact is that clouded leopards could benefit from captive breeding.

However, the 'mere' reproduction of animals is only the beginning. Animals in small populations – and zoo populations are inevitably smaller than pristine, ancestral populations – are liable to run into genetic problems, summarised as 'inbreeding', which reduce the fitness of individuals. Furthermore, the point of captive breeding is to retain a population that as closely as possible resembles the pristine, wild population. Only then, can captive breeding truly be called 'conservation breeding'. The aim, indeed, is to maintain animals that could be returned to the wild when necessary, and when the opportunity arises; with the hope that in a few centuries time it could become possible to return huge numbers to the wild. For such an ambition to be realised, conservation breeders must also seek to retain as much as possible of the genetic variation – the genetic potential – of the original, ancestral population.

In short, captive bred populations require *genetic management*. The ways in which this is achieved, and the underlying theory, are the subject of the next chapter.

FOUR

THE THEORY OF CONSERVATION BREEDING

The task is defined: to keep as many species as possible alive in captivity for as long as is necessary (which could well mean several centuries) in a state in which they are capable of returning to the wild. Most animals can be kept alive. Most can be induced to breed, at least from time to time. Yet all we have done so far is to stake out the ground. What we have seen, as Winston Churchill said in a somewhat different context, is 'the end of the beginning'. The problems that remain are those of biology, and in particular of genetics; and of logistics. These problems are the subject of this chapter. The first of these problems – one that pervades all further considerations – is that of numbers.

A MATTER OF NUMBERS

Noah, according to *Genesis*, saved all the creatures that were then alive simply by taking a pair of each kind into the ark. At first sight, this seems a reasonable strategy. A healthy young male and a healthy young female are enough to start a family; and if you can start a family, then (it seems) you can found an entire lineage, with generation following generation, and so on *ad infinitum*. There are some well-attested cases, too, of modern lineages beginning in just this way. Most famously, all the golden hamsters known in the world are the descendants of only one pregnant female who was found in the 1930s. The wild ancestors have never been found; and neither was the male who made her pregnant. Why is conservation breeding so difficult, then? It doesn't seem so hard, even among animals reduced to extreme rarity, to find one fertile male and one healthy female.

But Noah was blessed with divine intervention. In practice, in its absence, three different kinds of factor ensure that to begin a lineage that has a serious chance of surviving for more than a few generations you need a great many more founders than two. In some species, too, there are additional reasons why populations need to be large, in addition to the basic three.

The first factor is accident, which includes disease and the possibility of infertility. All of Noah's animals remained in miraculous good health. None of them broke legs, was eaten by another, or fell foul of infection. If key individuals drop out (and if you begin with only two, then they are both key individuals), then the future of the entire lineage is jeopardised. In real life, even in protected environments, such events are common-place. In some species (such as cheetahs) infertility is common; and until reproduction techniques are perfected infertility is as decisive as fatality. Indeed if you simulate an animal population with a computer model, showing accidents occurring at the kind of rate they occur in real life, then you can show that a population is extremely likely to go extinct within a dozen generations or so unless it contains at least 50 individuals at any one time.

Then there is the matter of *demographic stochasticity*. Demographic means 'pertaining to population'. Stochasticity means 'things that have to do with the nature of statistics'. The chief of these demographic glitches concerns sex ratio. In general, most animals produce an equal number of male offspring as females. By the same token, if you toss a coin a million times then you get roughly half-a-million heads and half-a-million tails. But if you toss a coin only half-a-dozen times, then you could easily get five heads and one tail, or even six tails or six heads. That does not happen very often, but it happens every now and again. Similarly, small populations of animals can produce an entire generation of offspring who are all of one sex. For slightly different demographic reasons (including selective mortality) the dusky seaside sparrow (a subspecies of seaside sparrow) in Florida was once reduced to six individuals, who were all male. (I will pick up on the story of the dusky seaside sparrow later.) For the first seven years of 'Operation Oryx' – the project to save the Arabian oryx, which I will discuss in Chapter 5 – all the calves born in the 'world herd' were male. Even after only limited experience, then, we can already see that what statistical theory predicts can happen *does* sometimes happen. It is difficult to say how many individuals are needed to avoid the problems of demographic stochasticity: it depends on so many other factors such as length of breeding life (can the parents breed again?) and fecundity (number of offspring). But again, we see that we would be far safer with dozens, or scores of individuals; and that some at least of Noah's pairs of animals would be bound to come to a sticky end.

We have seen, too, that some species simply do not breed well unless there is a crowd. Flamingos breed only in company, and pandas and perhaps sperm whales are among the creatures who possibly benefit from male competition. In very small populations, the stimulus for breeding may simply be too weak. However, the greatest reasons why populations need to be large in order to survive have to do with genetics. Here there are two main kinds of consideration. First – at the level of the individual –

it is important to avoid *inbreeding*, brought about by the mating of animals who are closely related to each other. Second – at the level of the population – it is vital for long-term success to conserve genetic *variation*.

As we will see, the number of individuals needed in a breeding population to avoid inbreeding varies enormously from species to species; but a rough, ball-park figure, to cover most eventualities, and sometimes regarded as a 'magic number', is 500. Five hundred is an enormous number to keep in captivity. Even in the wild, many populations are now below it. Note, too, that what counts in this context is not the total number of animals alive at any one time, for some of those may be infertile or senile; or there could be only one breeding female among 100 males – the majority of whom clearly would not have opportunity to breed; or half the animals could be identical twins of the other half, which means that matings between them would exacerbate inbreeding. What really matters is the number of animals capable of breeding; and indeed, to be really rigorous, the number capable of breeding who are not related to each other. This number is called the *effective population*. In reality (because some individuals in a population are likely to be infertile or senile, and many will be close relatives, and the sex ratio is not likely to be ideal) the effective population is liable to be less than the perceived population. Thus there are estimated to be between 34,000 and 54,000 elephants in Asia, which seems an enormous number. But only a few are breeding, or in a position to breed. So the effective population is much less than this.

Even 500, however, is not enough to maintain sufficient genetic variation to allow further evolution to continue in any one lineage. To preserve 500 or so individuals of any one kind must be seen, then, only as a holding operation. That core population must multiply again before any lineage can truly be considered to be back on the rails.

But we are running ahead of ourselves. Inbreeding (or the lack of it) and genetic variation are not the same phenomenon, though they are related. We should discuss each of them separately. First, though, we should establish the most basic of all concepts: the idea of the gene.

THE IDEA OF THE GENE

In the mid-19th century Charles Darwin solved, or at least outlined the answer to, the greatest of life's mysteries. He argued (as others had done before) that today's creatures had not been created in their present form; that they had evolved – 'descent with modification' – from ancestors that were different from their present-day descendants. He presented a plausible mechanism for this evolution: evolution by means of natural selection, which meant that in any one generation some individuals in any particular lineage would be better adapted to the prevailing conditions than others, and that these favoured individuals (the 'fittest' – meaning 'most apt') would be the ones to survive. He argued, furthermore, not simply that creatures change from generation to generation but that entire *species* could and did transform into new species; and, indeed, that any one ancestral species could branch to form many new, distinct species. That there could be change from generation to generation was not doubted in the mid-19th century; after all, cattle, sheep, horses, dogs, chickens, pigeons and all manner of plants had already been changed enormously in appearance and behaviour by the breeders of crops and livestock. But that *species* could change into new species, and that one species could branch into many – in other words, that 'descent with modification' could lead to radical change and to branching of lineages: those were, indeed, revelations. Many of Darwin's fellow scientists had as much difficulty with these ideas as did any theologian. Indeed the church embraced Darwin (for there is no reason why the Creator should not create by evolution if He chose to) while many excellent contemporary biologists protested throughout their lives.

But Darwin, odd though it may seem in retrospect, failed to solve a problem that now may seem to be crucial for a proper understanding of evolution. Though he wrestled with the issue all his life, he could never make head nor tail of heredity. He showed in general terms how the descendants of ancient apes could have evolved into us. Yet as he complained in his seminal *Origin of Species*:

. . . no-one can say why the same peculiarity in different individuals . . . is sometimes inherited and sometimes not so; why the child often reverts in certain characters to its grandfather or grandmother or other much more remote ancestor; why a peculiarity is often transmitted from one sex to both sexes, or to one sex alone, more commonly but not exclusively to the like sex.[1]

Curiously, the answer to Darwin's perennial puzzle was effectively solved within his own lifetime – indeed was being solved at the time he was writing *Origin of Species* in the 1850s – by an Austrian monk (working in what is now Czechoslovakia), Gregor Mendel. Mendel showed by breeding experiments how certain 'characters' (characteristics) in certain varieties of peas, such as wrinkliness of seeds and flower colour, were passed from generation to generation in an orderly fashion. Thus, he argued, each character must be conferred by the possession or non-possession of some 'factor' which the particular plant did or did not inherit from one or other of its parents. The notion that each character was determined by such a 'factor' was radically different from Darwin's notion – one shared by many of his contemporaries – that the inheritance of characters from generation to generation was like the mixing of inks.

It has often been suggested that if only Darwin had known of Mendel's work, then biology would have been advanced by 50 years. The reason he did not, it is suggested, is that Mendel published his seminal results with peas in an obscure Czech journal. However, Mendel's abbey at Brunn (now Brno) was by no means obscure. It was known for its excellent researches in plant breeding. Darwin was extremely widely read, and others were keen to bring key papers to his attention. It is at least possible, not to say quite likely, that he did know of Mendel's work. However, neither he, nor any of Mendel's close associates (who included some highly competent biologists) nor, it seems, Mendel himself, appreciated that the mechanism of heredity that appeared to apply to *some* characters of *some* peas would apply in principle in almost *all* characters in all creatures – animals, plants, fungi, and often in bacteria. Mendel knew even before he started his experiments that he would not get simple patterns of inheritance if he looked at other characters in other plants, or even in peas. He chose to measure the particular characters in the particular pea varieties because he knew from his previous observations that there *was* a simple pattern of some kind, and that measurement would yield instructive results. Many other characters, in peas as well as in other creatures, looked as if they were indeed conveyed from generation to generation as if by the random mixing of inks. Indeed there was absolutely no reason to suppose – in fact there was plenty of reason for not supposing – that the mechanisms of inheritance *were* the same for all characters, in all kinds of creature. Perhaps they had to be worked out *ad hoc*: character by character, species by species.

It was for such reasons, and not especially because Mendel was 'a simple monk' or because his workplace was 'obscure', that his work on heredity was effectively forgotten almost as soon as he published it in 1860, and that Mendel himself began increasingly to concern himself with the affairs of the abbey, rather than with plant breeding. His experiments were not re-discovered, indeed, until the early-20th century, by three different European biologists working independently, who realised its potentially resounding significance. One of them, William Bateson in England, coined the felicitous term 'gene' in place of Mendel's bland 'factor', to denote the unit of inheritance. The total complement of genes within any one individual is now called its *genome*.

Once the significance of Mendel's work was appreciated, the mysteries that had brought his researches to a stop, and which had continued to puzzle Darwin, fell away. Mendel himself perceived that any one gene (to use the modern term) could exist in more than one form (with each form now known as an *allele*. He also proposed that every creature contains each of its genes in duplicate: one copy from one parent, and one from the other. But each parent passed on only one of the duplicate pair to its offspring. In practice, any one offspring might inherit the same form (allele) of any one gene from each of its parents, in which case it is said (in modern parlance) to be *homozygous* for that gene (or for the character conferred by that gene); or that it could, in principle, inherit a different allele of any one gene from each of its parents, in which case it would be *heterozygous* for that gene.

In practice, Mendel found that if he crossed pure-breeding round-seeded peas with pure-breeding wrinkled-seeded peas then all the offspring would have round seeds. He did not throw up his hands in despair, however. Indeed he did the exact opposite. With remarkable percipience (he really was an excellent scientist) he argued as follows. Each of the offspring must inherit a round-seeded allele (again, in modern parlance) from one parent, and a wrinkled-seeded allele from the other. But, he said, the round-seeded allele is *dominant* (and the wrinkled-seeded one is *recessive*). So the offspring all have round seeds.

However, in the language of modern genetics, the round-seeded offsrping are, in reality, heterozygous for the particular gene of seed-coat structure. When heterozygous creatures become parents, then they have an equal chance of passing on either one of the two alleles. So if the offspring of two peas that were heterozygous for wrinkled seeds or round seeds were crossed, then some of their offspring would inherit a round-seeded allele from each of the parents, and they would be homozygous, pure-breeding round-seeded plants; some would inherit a wrinkled-seed allele from each parent, and be homozygous, pure-breeding wrinkled-seed plants; and some would inherit a round-seeded allele from one parent and a wrinkled-seeded allele from the other, and they would be heterozygous, round-seeded plants that did not 'breed true' (because they in turn could pass on either a wrinkled- or a round-seeded allele to their offspring).

Mendel was able to infer all this because he crossed a great many peas, and when you do produce a great many, then simple patterns of inheritance begin to emerge. Thus, when heterozygous peas are crossed with each other, one-quarter of the offspring inherit a round-seeded allele from both parents; half inherit a round-seeded allele from one parent and a wrinkled-seeded one from the other; and a quarter inherit a wrinkled-seeded allele from both parents. The quarter that inherit two round-seeded alleles of course produce round seeds; the half that have one round-seeded and one wrinkled-seeded allele also produce round seeds, because the round-seeded allele is dominant; and the quarter with two wrinkled-seeded alleles have wrinkled seeds. Thus the ratio of round-seeded to wrinkled-seeded offspring is three to one; and if you are Gregor Mendel, and do further crossing experiments with other combinations, you can infer the notions – in modern terminology – of genes, alleles, heterozygosity and homozygosity, and of dominance and recessiveness.

Note just three points from this most elementary example. First, the physical appearance of a plant or animal – its phenotype – may not give you an accurate picture of the kinds of genes it contains – its genotype. Thus, the round-seeded plants could be either homozygous or heterozygous; you just cannot tell by looking at them which is the case. But when round-seeded plants are mated with each other, you can see whether they were heterozygous or homozygous by looking at their offspring. If all the offspring are round-seeded, then the parents must have been homozygous for the round-seeded allele. But if some are wrinkled-seeded, then the parents must both have been heterozygous for that gene, for only then could some of the offspring have inherited two wrinkled-seeded alleles, which they have to do if they are to have wrinkled seeds. However, you do know that all plants with wrinkled seeds are homozygous for that gene, because if they possessed a round-seeded gene then they would indeed have round seeds, because the round-seeded allele is dominant.

The second point is that this example really is simple: Mendel chose to work on characters that he knew (because he was an extremely knowledgeable plant breeder, apart from being a scientist) would give simple patterns of inheritance. Even so, however, once we introduce the essential concept of dominance and recessiveness, we see how very complicated real life really is. Round-seeded peas can produce wrinkled-seeded offspring – if they themselves are heterozygous. Wrinkled-seeded peas cannot produce round-seeded offspring, because if they were heterozygous they would not be wrinkled-seeded.

Note, finally, however, that the simple patterns of inheritance from which Mendel inferred so much are not evident if you look at only a few offspring of a few parents. We are back to the matter of stochasticity. With a huge sample, it will indeed emerge that the round-seeded homozygotes and heterozygotes outnumber the wrinkled-seeded homozygotes in a ratio of three to one. But it is chance that determines whether any one plant inherits a round-seeded or a wrinkled-seeded allele from a heterozygous parent. By chance, the first six offspring, or even the first hundred offspring, might all inherit wrinkled-seeded genes alone. Then all the offspring would be wrinkled-seeded. In slow-breeding animals, such as human beings, there are never enough offspring in any one family to see the patterns. In humans, the allele for brown eyes is dominant over the allele for blue eyes. But we all know that in real families the children of brown-eyed parents may all have blue eyes. No wonder Darwin could not work out the basis of heredity. He was a broad thinker; his strength lay in looking at all the features of all the creatures he came across. But if you do that, you cannot see the patterns. Only by minute study of particular characters in particular plants or animals can you see the principles. Mendel's brand of genius lay in his pernicketiness.

Yet real life has many more complications to offer. First, most visible characters are not determined by only one gene. Most are polygenic: that is, they are determined by several genes working in concert. Once you start working through examples of heredity with such characters, the patterns of heredity rapidly become mind-blowing. Then again, in domestic animals and plants (and Mendel was studying cultivated peas) each gene may exist as only one or a few different alleles. But in wild populations any one gene may be highly *polymorphic*; that is, it may exist in many different alleles. You may like to work out the patterns of inheritance that you would get for a hypothetical character in a hypothetical population that was determined by three genes, each of which could exist in three alleles. Or then again, you may not.

Yet another complication is that of partial or incomplete dominance. The round-seeded allele dominates the wrinkled-seeded allele totally, to be sure. But this is not always the way. For example, some plants have alleles that produce red flowers, and alleles that produce white flowers. But if the two are crossed, the offspring have pink flowers. It is this kind of example that gives the (false) impression that inheritance is a matter of mixing inks. Then again, most genes are *pleiotropic*: that is, they affect more than one character. Often the two or more different characters that are affected by the same gene are quite unrelated to each other; a gene that affects lung function might also influence coat colour and length of tail, for example.

The final important factor that complicates patterns of heredity is *linkage*. One of Mendel's seminal findings was that different characters in the same plant were, as a rule, inherited entirely independently. Thus he looked at the inheritance of seed colour in peas, as well as seed form: yellowness or green-ness. He found that seeds could be wrinkled and yellow or wrinkled and green, or round and yellow or round and green. Form and colour had nothing to do with each other.

But this is not always the case. Some particular characters are almost invariably inherited together with other particular characters. Sometimes pleiotropy is the cause: the same gene causing two quite different effects. But often the explanation is that two different genes are physically linked, so that they are passed on together.

The point here is that genes in practice are packaged in *chromosomes*. Each chromosome is a string of genes, held in place (at least during cell division) by a core of proteins. Each gene occupies a particular place on its allotted chromosome, known as its *locus*. In general, each kind of animal has a characteristic number of chromosomes: humans have 46, horses have 42, and so on. The number seems to be a matter of chance: just the way evolution worked out. In general, though (the exceptions need not trouble us), all animals have an even number of chromosomes. In fact they have two sets; one set inherited from the mother, and one from the father. The two sets between them contain the entire genome (again with exceptions which need not trouble us).

Each cell in each animal, in general, carries a complete double set of chromosomes. During cell division, each chromosome splits lengthways down the middle, and then each half duplicates itself, to give four sets – a process known as *mitosis*; and then two sets migrate into each daughter cell to produce two facsimiles of the parent cell. This can be observed under the light microscope.

But during the formation of *gametes* (eggs or sperm) a different form of chromosome division takes place, known as *meiosis*. The initial division takes place, to produce four sets of chromosomes; but then only one set migrates into each of four gametes. Hence each gamete contains half the genome of its parent. Fusion of two gametes produces a brand new, mixed genome. The physical facts, as observed down the microscope, precisely (one is almost inclined to say miraculously) support Mendel's interpretations. But the behaviour of chromosomes was not observed until the late 19th century; so Mendel was working without this background knowledge.

From this description, however, it might seem that the gamete contains a rather badly mixed selection of genes. One might envisage that the gamete finishes up with some entire chromosomes that originally came from its parent's mother (that is, from the gamete's grandmother), and some that originally came from its parent's father (the gamete's grandfather). But during meiosis, another process supervenes. Each maternal chromosome first lines up with its equivalent paternal chromosome. Then each one splits lengthways, as in mitosis – to give a collection of four split chromosomes, all lying side by side. These then jumble up, in a process known as *crossing over*, to produce some entirely novel chromosomes, each containing some genes from the maternal chromosome, and some from the paternal. Because of crossing over, genes may be inherited independently even if they occur on the same chromosome. But if two genes occupi loci that are close to each other on the same chromosome, then they are unlikely to be separated by the jumbling process of crossing over. Hence they are passed on in tandem. Hence, linkage. If two characters are invariably inherited in tandem, then we might suspect pleiotropy. But if the characters are occasionally inherited independently (because crossing over will sometimes separate them) then linkage is the probable explanation.

One final note before we pass on. For meiosis to succeed, and for gametes to be produced that contain a proper assortment of grandmother and grandfather genes, it is essential that each of its parent's maternally-derived chromosomes should be able to line up with its paternally-derived partner. Otherwise crossing over cannot take place, and the whole meiotic process aborts. This is why two different kinds of animals generally cannot mate to produce fertile offspring unless they themselves have the same number and arrangement of chromosomes; for the offspring (assuming they succeed in producing one at all) will not have the necessary matching pairs of chromosomes. A difference in chromosome number and arrangement is not the only factor that separates different kinds of animal. But it is one of them; and, as we will see, in the context of conservation breeding it can be important.

There are other complications in heredity, though for present purposes they are not of huge significance. It is clear, though, that even a few conceptually simple complications produce enormously complex, indeed bewildering, end results. The virtual dismissal of Mendel's early results was not at all surprising. It was not simply a piece of carelessness. His simple pea ratios really did not seem to have much to say about the erratic inheritance patterns of life in general. The biologists who later perceived their significance showed remarkable insight.

We will return to the question of the relationship between Mendel's genetics and Darwin's theories of evolution. First, though, we should discuss the first of the key genetic issues that can so severely influence the fate of animals: that of inbreeding.

INBREEDING

Any gene in any population can exist in more than one version, each version being known as an allele. Genes (as I will explore in more detail later) operate by determining what kinds of proteins are produced, and proteins are highly functional molecules, which provide much of the body structure (including the contractile parts of the muscles, and much of the cell membranes), *and* act as enzymes which are the catalysts that control most of the body's functions. Some proteins also act as hormones. Thus differences in the proteins that the genes produce profoundly affect the structure and function of the whole body, and hence the *kind* of creature that is produced.

Many genes, to be sure, are non-functional; they are 'switched off' throughout the creature's life. Many others function only for brief periods, before and after which they are switched off; including, for example, many of the genes that operate within the foetus. All the cells of the body in any one creature have the same complement of genes, but they may differ enormously from each other in form and function because in different kinds of cell different permutations of those genes are functional. Thus many genes are functioning in liver cells which are switched off in muscle cells, and vice versa. Some genes function in all cells throughout the creature's life.

Because genes play such a crucial role in the life of each creature, it matters which version – allele – of any one gene is inherited. Admittedly it does not always matter. If the gene happens to be one that is permanently switched off, then it does not matter at all. It does not matter, particularly, either, whether a European inherits an allele that will give him or her blue eyes, or one that will give brown eyes – though small changes in the colour of some external feature could profoundly affect the success of some creature that relied on visual features to attract mates or to avoid predators. In humans (as in most animals and plants) many alleles affect height and build. If a person lives in an extreme climate without modern comforts, then genes that affect stature matter; Eskimos are better off being short and stocky, while Sudanese shed surplus heat more easily if they are tall and slender. In Europe it does not matter very much. Tallness is generally considered an asset among men but many short men (such as Napoleon) do very well. By contrast, the alleles that affect the structure of haemoglobin, the red pigment that carries oxygen in the blood, do not produce directly visible effects (unless you examine the blood under a microscope) but their effects matter a very great deal. There are many alleles that produce anomalies in haemoglobin structure which result in various forms of inherited anaemia: sickle cell anaemia of Africa, and various thalassaemias in the Mediterranean and South-east Asia.

Genes are not perfectly stable. From time to time they change – mutate. Mutations are extremely unlikely to occur in all individuals of any one population at the same time, which is why, at any one time, the individuals in any one population at any one time are likely to contain several or many versions (alleles) of any one gene.

Mutations, producing brand-new genes with brand-new effects, are the feedstuff of evolutionary change. If mutation did not occur, then we would still have the same genes as our bacterial ancestors who lived 2 billion years ago; in other words, we would still be bacteria. But mutations occur by chance. Functional organisms, whether bacteria or human beings, are delicate and well balanced, and any one mutation is as likely to produce improvement as is, say, a random blow aimed at a television. Most mutations, to be sure, probably have very little effect, for most are small: perhaps just a small shift in some redundant part of some protein. Some have hugely deleterious effects, and their unfortunate inheritors are instantly wiped out, and so the harmful genes are not passed on. Some mutations – a very small minority – in the fullness of time, other changes having been made, actually increase fitness, and they are favoured by natural selection and contribute to evolutionary change.

There is a catch, however – one brought about by the phenomenon that Mendel noted: that of dominance and recessiveness. If a creature inherits an allele that produces deleterious effects from only one of its parents; and if it inherits a normal allele of that gene from its other parent; and if the deleterious allele is recessive – then the effects of that deleterious gene will be masked, its possessor will not suffer its ill effects, and will live to pass on that deleterious gene to the next generation. The heterozygous possessor of a deleterious recessive gene is said to be a *carrier*. In practice, natural selection will tend to eliminate deleterious alleles that are dominant, for every creature that possessed such an allele would be obviously disadvantaged. But it will not so easily eliminate deleterious alleles that are recessive, because their effects will not be seen among the heterozygous carriers. We can put this another way, and say that natural selection favours recessiveness in deleterious alleles.

We have seen, though, that recessive alleles do make themselves felt if inherited from both parents. If a heterozygous carrier of a deleterious allele mates with another heterozygous carrier, then one quarter of the offspring (on balance) will inherit a normal allele from both parents, and they – and their offspring – will be totally free of the deleterious gene. One half of the offspring will inherit a normal allele from one parent and a deleterious allele from the other, and will in their turn be carriers: unaffected themselves, but capable of passing on the deleterious allele. And one quarter will inherit the deleterious allele from both parents, and they will be adversely affected.

Several points are worth noting. First it is clear that the carriers are bound to out-number the homozygous sufferers. For example, it seems that one in 20 North European people – an extraordinarily high proportion! – carry the allele that causes cystic fibrosis, when inherited in double dose. Each such carrier has a one-in-20 chance of marrying another carrier, so the chances of two heterozygotes marrying are one in 400. One in four children of such a marriage will have the disease. So we would expect to find that one in 1,600 North European infants is born with cystic fibrosis. That is precisely what we do find. If a particular deleterious allele was carried by only one in 50 people (which still seems common enough) then the disease it causes would appear once in only 10,000 infants. And so on.

Note, too, then, that deleterious alleles are not easily eliminated. Enthusiasts for eugenics (who seek to 'improve' the genetic content of the human species) have sometimes suggested that the unfortunate sufferers from genetic diseases ought not to be helped to lead normal lives by modern medicine, because they will then be helped to pass on their deleterious genes. In practice, help to the sufferers makes very little difference. The unfortunate sufferers from cystic fibrosis are unlikely to breed in any case, because at present there is no 'cure'. If they could, they would be unlikely to choose to do so. But if they were able to breed, and did choose to do so, then their one-in-1,600 contribution to the total frequency of the gene would make very little difference. To eliminate the gene for cystic fibrosis, the eugenicist would have to examine minutely, and then sterilise, one in 20 of the North European population. In practice, too, cystic fibrosis is only one of more than 3,000 diseases identified in human beings that are known to be conferred by single genes. This implies (once you do the arithmetic) that each human being carries an average of about six recessive deleterious alleles which would cause frank disease if paired with another like itself. To purge the human race of its potentially lethal genes, enthusiastic eugenicists would have to eliminate every one of us, and preferably six times over. Including themselves.

Note further, though, that frank diseases such as cystic fibrosis are only extreme examples. Alleles may fairly be said to be deleterious in a competitive world even if their ill effects are much less obvious. Thus we know now that most of the diseases that human beings suffer from, from coronary heart disease to infections, have some genetic basis. Many of these more covertly deleterious genes do not make themselves felt until their possessor is past reproductive age (coronary heart disease mainly affects the middle-aged and old) and so they are even harder to eliminate. But if such genes produced effects early in life, and if people lived without the benefit of modern medicine, then their possessors would be disadvantaged. In principle, though, we can envisage any degree of disadvantage. Koalas, for example, have enzymes in their livers which help them to detoxify the otherwise lethal eucalyptus leaves on which they feed. A koala with a less than normally functional detoxifying enzyme would be disadvantaged, and in a competitive environment would tend to leave fewer offspring.

But what has all this to do with inbreeding? Let us again take cystic fibrosis as an example. A carrier has a one-in-20 chance of marrying another carrier if he or she mates at random with some other member of the North European population. Suppose, though, that a carrier decides to marry his sister. Even if he chooses a healthy sister (which is most likely) there is still a two-out-of-three chance that she, too, is a carrier. If she is not a carrier (a one-in-three chance) then all the offspring will be healthy, but half of them will be carriers. If she is a carrier (a two-in-three chance) then a quarter will be healthy, a half will be carriers, and a quarter will have the disease.

In most societies brother-sister matings are frowned upon and rare; but in many societies and sections of society matings between near relatives are either forced (because the population lives in some isolated settlement) or encouraged (for example among aristocrats, who do not want their lines to be corrupted with commoners 'blood'). In such societies, single-gene disorders can be extremely common. Haemophilia, schizophrenia, and porphyria are among the clear-cut gene-based pathologies that are known to have run through various European royal houses.

As we have said, though, there are infinite shades of deleteriousness. Frank disease is only the extreme example. Among animals, a common consequence of excessive inbreeding is infertility. As I will explore later, it is possible now to look directly at the structure of genes, and thus to assess whether and to what extent an individual is homozygous – which implies that they are inbred. Such studies have shown that cheetahs in particular are extremely homozygous: all the cheetahs in Africa are genetically very similar to all the other cheetahs. The explanation (we do not know that it is true, but it is the only one that makes sense) is that some time in the past the population of cheetahs must have been severely reduced, so that all present-day cheetahs are descendants of just a few offspring; in other words (as the expression is) cheetahs in the past went through a genetic bottleneck. In effect, then, all modern cheetahs are fairly inbred, even when they live in the wild and mate randomly. Male cheetahs in particular have a notoriously high level of infertility – a fact that has greatly complicated captive-breeding programmes. If lack of fertility were closely correlated with lack of libido or personal charisma this would perhaps be less important. It has often transpired, however, in several species (including gorillas) that the dominant male, which does most of the mating, is sterile. These things are sent to try us.

You may well object, however. Cheetahs are inbred, to be sure, but they exist. All the golden hamsters in the world are the offspring of the only wild female ever found. Domestic cattle and dogs may be highly inbred – deliberately so, to keep them uniform; and laboratory mice are *extremely* inbred. Two observations. First, despite all the above remarks, most inbred animals turn out to be less inbred than is at first apparent, when you look very closely. Even laboratory mice, which are zealously inbred over many generations, have often turned out to be embarrassingly variable in the minutiae of their body chemistry. The pregnant female golden hamster who founded the later lineage was herself highly heterozygous – as attested by the fact that her descendants are actually rather variable.

Second, and more importantly, the existing in-bred lines are the lucky ones. It is theoretically possible, after all, for an animal to contain no significantly deleterious alleles at all. We said that humans on average contain about six deleterious alleles. Statistically, there must be some who contain none. If these individuals were to be selected, and bred together, they could presumably produce dynasties that could be highly inbred, but yet be healthy. We know, though, that for every existing strain of laboratory mouse there have been hundreds of failures: lines that bred for a few generations and then collapsed through sterility or some frank pathology. The history of agriculture shows how often they have been obliged to re-vivify their stocks by importing unrelated animals from outside (and perhaps, indeed, from foreign countries). Nobody would have been surprised if golden hamsters had petered out after a few generations. Wild cheetahs evidently survived their bottleneck. We do not know how many other kinds of creature failed to survive similar bottlenecks.

Inbreeding, then, is to be avoided. Indeed, geneticists speak generally of *inbreeding depression*, to denote the general lack of fitness that results from it: anything from a slight loss of competitive edge, to frank disease. If geneticists know the ancestry of each animal in a breeding population, and know who has sired and given birth to whom, then they can work out the degree of inbreeding that must have occurred within this population; and can express this as a precise mathematical expression known as the *inbreeding coefficient*. An inbreeding coefficient of zero represents no inbreeding; while a coefficient of 1.0 denotes complete homozygosity. Clearly, in general, the aim is to keep the inbreeding coefficient as low as possible; and because it can be quantified, the breeding strategies that will keep it low can also be worked out in precise terms.

By contrast, if two animals of the same general type mate, so that their offspring are extremely heterozygous, then these offspring are sometimes especially vigorous. Darwin noted this, and indeed coined the expression 'hybrid vigour' to describe it. I stress the 'sometimes', though, because, as we will see later, the offspring of genetically different types are not always especially favoured – hybridisation can definitely be overdone – even though they clearly benefit from heterozygosity. The vigorous hybrids are not always quite as vigorous as they seem, either. Mules, for example, produced by crossing horses and asses, are extremely 'vigorous' in the sense that they have amazing stamina and get by on a meagre diet. But they are also sexually sterile, which of course means that their own particular form of 'vigour' cannot be passed on.

However, the minimisation of inbreeding (and the reduction of the inbreeding coefficient) is only a part of what is necessary. The other part is to maximise genetic variation within each population.

GENETIC VARIATION

What, first of all, is meant by 'genetic variation'? It is, after all, an odd concept. We talk of genetic variation within a breeding population. Yet animals that are in the same breeding population are of the same kind, and animals of the same kind have the same genes. Don't they? In practice no, of course, they do not. Animals in the same breeding population do have the same number and arrangement of chromosomes. (In fact in mammals, birds, and many other groups of animals the females have a different arrangement from the males. But the differences are consistent, so the generalisation that animals of the same kind have the same kind of chromosomes is good enough.) In practice, too, each chromosome in each individual has the same number of loci as each equivalent chromosome in all the other individuals.

But each gene within each locus may exist in one of several versions – alleles. So two individuals in the same sexually-reproducing breeding population will have the same number and arrangement of loci (sexual variations aside) but unless they are identical twins they will each have a different assortment of alleles. To be sure, within any one population some genes will exist only in one form (allele), and that particular allele is then said to be *fixed*. But other genes – especially those associated with disease resistance – may exist in a great many different forms; and such alleles are said to be highly *polymorphic*. The total range of different alleles within any one breeding population is called the *gene pool*. 'Genetic variation' is a way of expressing the degree of variety within the gene pool.

It is at this point that we can see the true connection between the ideas of Mendel and those of Darwin. To be sure, when Mendel's ideas were first rediscovered in the early 20th century, they were thought to be incompatible with Darwin's. After all, Darwin's theory of evolution by means of natural selection seemed to demand gradual change; but it seemed that Mendel's genes caused sudden, all-or-nothing change. For a time, the growing band of geneticists were at loggerheads with the Darwinians. That the two sets of ideas were compatible in principle became apparent as the true complexities of genetics were revealed. If a character is polygenic (as most are), and if most mutations are small in their effects (as they are), then gradual change is certainly possible.

But the final fusion of the two sets of ideas – what Julian Huxley called 'the modern synthesis' – was not made until the 1940s.[2] Then biologists came to see natural selection as a shift in the composition of gene pools. Thus, if any one individual within a population possesses a gene that confers an advantage, then it will tend, on average, to leave more living offspring than one that lacks that gene. Those offspring themselves are liable to carry the advantageous gene. Populations of animals do not grow indefinitely; and if the population stays the same size, then the individuals that possess the gene will spread at the expense of those that do not. Hence the advantageous gene becomes more common, while less advantageous genes become rarer. Hence, over time, natural selection changes the composition of the gene pool. This view of evolutionary change, combining Mendel's ideas with Darwin's is called neo-Darwinism, and, in essence, it represents the modern orthodoxy of evolutionary thinking.

That is not the end of neo-Darwinism, however, because (as Darwin himself appreciated) natural selection is not the only cause of evolutionary change. Time and chance play an enormous part as well. Thus, a gene pool tends to gain new alleles – increase in variation – as time passes, simply because new mutations occur. Natural selection may of course act to increase the frequency of the new mutation, as we have discussed. Or it may tend to diminish it. But the effect of the mutation may be weak (at first!), and the mutant gene may be recessive; in which case the new mutant may hang around, contributing to variation, but not significantly affecting the lives of the animals. At some point in the future, however, such a mutant may turn out to be useful (or disadvantageous), in which case it will become the target of natural selection. But the initial variation on which natural selection acts arises by chance; and natural selection cannot act upon what does not yet exist.

Second, gene pools lose alleles by *genetic drift*. This happens in two main ways. First, any allele that is rare is (by definition) carried by only a few individuals – or indeed, in the end, only by one. If such an individual dies before it reproduces, then the rare allele dies with it. But there is a more general mechanism of genetic loss than this. Each animal, as we have seen, passes on only half its genome to each of its offspring. Unless an individual has a great many offspring, like a fly, then there is fair chance that it will fail to pass on at least some of its alleles to its offspring. Hence, as each generation passes, there is likely to be a loss of alleles. This is avoided only if the population is extremely large, or if it contains no rare alleles. Only in such populations is there a good chance that every allele will be passed on. Only these, then, can truly be considered 'viable'. Those that are continuing to drift may drift to extinction.

Thus, in a typical, wild population, neo-Darwinists envisage a steady expansion of the gene pool due to mutation, and a steady reduction due to the two main causes of genetic drift. (Note here that we talk of expanding or contracting the gene pool; not necessarily of expanding or contracting the population itself. The gene pool can expand without increase in individuals if those individuals become more heterozygous; and vice versa.) But superimposed on this shifting tide of alleles within the gene pool is natural selection; knocking out the less advantageous genes, and so making room for expansion of the more advantageous. Taken all in all, this general picture seems to explain the events of nature very well, and to provide a good conceptual platform on which to found breeding strategies.

Why, though, is genetic variation desirable? What does it matter if all the animals are uniform? The answers are already obvious but they are worth talking through. First, if all the individuals are genetically identical, then they will also be identical (or extremely similar) in what they can do, and what they are adapted to. If one individual is especially susceptible to drought, then they will all be susceptible to drought, and if it fails to rain, then they all die. Of more general relevance is that any strain of pathogen that kills any one individual can kill them all. Cheetahs, which are extremely uniform, are also extremely susceptible to epidemic.

Second, we have seen that natural selection is essentially a destructive force. It does not create useful changes; it simply selects among the variations that are available, and knocks out those that confer least advantage. Natural selection clearly cannot operate *unless* there is variation to act upon – and the greater the choice, in general, the better. But although natural selection is not the only force that operates in evolution – for time and chance play an enormous part – it is the only known force that leads to *adaptive* change. Without innate variation, then, lineages cannot continue to adapt and re-adapt to their ever-changing circumstances as they have done through all the millions of years of the past. In fact, if genetic drift is operating rapidly, then populations can in principle lose valuable adaptations, which accelerates their extinction.

Finally, there is in practice in animal populations a close link between the variation within the gene pool as a whole, and the degree of homozygosity within each individual; and excessive homozygosity, as we have seen, is the cause of inbreeding depression. At an intuitive level, the reason is obvious. If all the animals in the population contain the same alleles, then every offspring is bound to inherit the same alleles from its father as it does from its mother. Variation in the gene pool is, in general, one of the best antidotes to homozygosity in individuals.

Just to be pernickety, it happens to be the case that the degree of homozygosity in the individuals is not necessarily related to the amount of variation in the population as a whole. Wheat plants, for example, are naturally inbred: the pollen generally fertilises the ova on the same plant. This means that individual wheat plants, even in the wild, are extremely homozygous. However, within a population of wheat you may find unrelated individuals, each of which is inbreeding. So each individual is homozygous, but the population as a whole may be variable. By contrast, crop breeders commonly cross highly homozygous (and uniform) lines of plants such as maize, with highly inbred lines of an unrelated strain of maize. The individual offspring that result from such crosses are all extremely heterozygous; but because they are all the offspring of extremely uniform parents, they too, as a group, are extremely uniform. Such plants are the 'F1 hybrids' (where 'F1' means 'first generation') now so beloved of gardeners.

Even in animals, we may see highly heterozygous populations that are uniform: for example, the offspring of Friesian cow mothers and Hereford bull fathers, which are commonly raised for beef in the UK. But in wild populations, there is no farmer to control mating, and animals do not normally inbreed as wheat does. So in wild animal populations, where the males and females take steps (as they commonly do) not to mate with their close relatives, the degree of heterozygosity in individuals and the degree of variation in the gene pool as a whole are very closely correlated.

From all the above remarks, both on genetic variation and on inbreeding, our opening comments become obvious: the chances of survival are enormously enhanced by large numbers. For example, each individual animal can carry only two alleles of any one gene: one on the maternal chromosome; one on the equivalent paternal chromosome. Two heterozygous individuals that were unrelated could, at most, carry four alleles of the same gene. Clearly, though, the chances of any two individuals carrying as many as four alleles of more than a few genes are remote indeed. If a population is truly to contain a good variety of alleles of a high proportion of its genes then it must contain a great many individuals.

Second, we can see how as populations diminish, the problems multiply, so that in the end – even though there may seem to be quite a few individuals remaining – the population may go into a tailspin. For example, if a rare allele is contained within three individuals, and one of those fails to pass the allele on – that is a diminution, but the rare allele none the less lives to fight another day. But if the allele is contained in only one individual, and it is not passed on, then it disappears forever. Thus genetic drift bites hard when populations are small. Island populations, suddenly cut off from the mainland, may change markedly in a few years through genetic drift alone.

As the gene pool is diminished by genetic drift, so, obviously, the remaining animals become more uniform. Thus homozygosity increases. In addition, it is bound to be the case in small populations that only a few females are breeding. So a high proportion of the ensuing generation are siblings. Homozygosity increases yet again.

But then, as a population becomes really small, demographic stochasticity may start to bite. Suddenly we find we have a generation all of one sex. Or we find that only one or two females are breeding – and then the inbreeding problems become very great indeed. Finally, if accident does not wipe them out, the remaining creatures will peter out simply through inbreeding depression manifesting as infertility. Unless, like the cheetah, they are very lucky indeed.

Pressure on space in zoos is such that we cannot hope to save enough individuals of any one species to enable natural selection to continue as it has in the past. The best we can hope for is a holding operation. The immediate aim is to retain enough variation to avoid inbreeding depression. The longer-term aim is to keep enough variation so that when the animals are returned to the wild some individuals at least will pull through. Of course, the initial elimination in the wild of the individuals who are not well adapted is natural selection of a sort: but there will not be sufficient variation in the short term to enable further adaptation to take place after the initial purge. What survives in the wild in the short term will be what there is. As numbers build up, however, to thousands and tens of thousands, then natural selection will truly begin again to remould the lineage to the requirements of the environment.

Even to achieve the holding operation is not easy, however. Zoos cannot keep huge numbers of each creature, no matter how much they cooperate. The requirement for the present, then, is to maintain as much genetic variation as possible within populations that are inevitably smaller than ideal, this being an aspect of what is generally termed small population management. To maintain genetic variation and avoid inbreeding in small populations requires orchestrated breeding strategy. This we must now discuss.

MAINTAINING VARIATION: THE TARGET

So far, we have talked in broad generalisations. Breeding populations should be 'as big as possible'. In general, 500 of each kind is a good ball-park figure. But in zoos, space is at a premium. If zoo conservationists breed more of a particular species than are strictly needed, then they will inevitably leave some other creature out in the cold. But if they breed too few of any one kind, then their efforts are wasted, because inbreeding and extinction will begin to seem inevitable. It is important, then, to tighten up on the figures.

It is clear, too, that some loss of genetic variation is inevitable with each generation, just because each breeding animal passes on only half of its genes to each of its offspring. To avoid this entirely, populations must be far larger than is realistic. Again, then, captive breeders must acknowledge that ideals are impossible, and agree upon a target that can be achieved.

Finally, as genes are lost when the animals breed, the total proportion that is lost increases as the generations pass. I said in chapter 2 that the ideal is to keep captive animals in a viable state for 500 to 1,000 years, for it will not be until then that (perhaps) significant space will become available in the wild. But it is far harder to maintain genetic variation for 500 years – which for some animals might mean 500 generations, with 500 chances to lose genes – than for some shorter time. Again, it seems wise to seek some more realistic target. In practice, then, conservation breeders such as Dr Tom Foose, conservation coordinator of the American Association of Zoological Parks and Aquaria, have agreed realistic targets, and, from these, have been able to calculate just how many individuals of each kind of animal must be kept in order to achieve them.[3]

First, they have shown by genetic theory and computer modelling that it is very much easier to conserve 90 per cent of existing genetic variation than to conserve 100 per cent. Ninety per cent is a lot; it ought to be enough to provide populations that are viable. Second, conservationists have observed that some of the new technologies that I will discuss in chapter 6 – the freezing of gametes and embryos, the transfer of embryos – are developing rapidly; and that these, well applied, can solve many of the problems by enabling us to keep genes 'on ice', without maintaining entire, expensive beasts. In 200 years, the scientists reason, these techniques should be very advanced.

These two lines of thought, then, have defined the target now agreed by most zoo conservationists: to maintain 90 per cent of existing genetic variation for 200 years. But how can this target be achieved?

The means to variation: founders

To begin with, breeders can maintain only those genes that they have access to: only those that are contained within the *founders* of their captive population. It follows, then, in general, that captive populations should be initiated with as many founders as possible. Clearly, too, each of those founders should ideally be as heterozygous as possible, and each one should be unrelated to all the others. That way, the founding gene pool is as large as possible.

In practice, it transpires that two individuals (male and female) that are taken from different parts of their natural range so that they are likely to be unrelated could contain as much as 75 per cent of the total range of alleles found in the entire population. That is more than might be expected. But 75 per cent is achieved only with an ideal pair; and it still falls short of the desired 90 per cent. Theory suggests that to encompass 90 per cent of the variation in a population you need to take at least six unrelated individuals. Six may seem an encouragingly modest number, but in practice there are snags. The case for beginning captive-breeding programmes is strongest when wild populations are already rare. But if they are rare, then it becomes a big decision to take even six from the wild – especially six young healthy animals from different parts of the range. This, clearly, would compromise the wild population even more. In addition, for many species it is no longer possible for legal reasons to take any animals from the wild. Thus, captive-breeding programmes generally rely heavily on founders who are already in zoos. One snag is that the ancestry of zoo animals is sometimes unknown (because some zoos in the past did not keep good records); though it is clear that many zoo animals are closely related to other zoo animals, either because they were both born in the same zoo from the same parents, or because they were initially captured from the wild as a litter.

In practice, as captive breeding improves and wild populations are increasingly threatened, zoo conservationists are more and more inclined to begin captive-breeding programmes long before the numbers in the wild have fallen to disastrous levels, which makes it easier to begin with a reasonable number of founders. Sometimes, too, the wild population is so endangered and so rare that the only sensible decision is to bring the whole lot into captivity. As I will discuss in the next chapter, this has been done in North America for the California condor, the red wolf, and the black-footed ferret. In general, though, as we will again see in the next chapter, different captive zoo populations show different degrees of compromise. Some easily achieve the target of six unrelated founders, and some do not.

However many founders there are, their influence lingers for many generations. Thus, the number of individuals kept in a captive population will, in the end, depend on the number of spaces available for them. Suppose the final population achieved, by all cooperating zoos, is 200 animals. If the population began with four founders, then that 200 will contain genes derived from only those four; but if it began with six, then it will contain genes from six – which should mean a greater variety of alleles, if the original founders were not themselves inbred. As the generations pass, a few alleles will be lost by genetic drift, even though the number of individuals may remain at 200. But in any generation, the population that had six founders should have a greater genetic variation than the one that began with four. Of course, if the four-founder population had been increased initially to 500, then it would lose a smaller number of its alleles with each generation; so that perhaps after a few generations a 500-individual four-founder population might encompass as much genetic variation as a 200-individual six-founder generation. But if other things remain equal – the same-sized population after the same number of generations – the six-founder population will remain more varied for a very long time.

Rapid multiplication

We have seen that Noah's approach to conservation left much to be desired. But God did offer one piece of excellent advice. As the creatures came from the ark He bade them to 'breed abundantly in the earth, and be fruitful, and multiply upon the earth' (*Genesis* 9.17). Absolutely. If you start with only a few founders (as seems inevitable) then it is essential that each one should in the short term breed as abundantly as possible. This is by far the best way to ensure that as many as possible of the founders' alleles are passed on. This is where modern techniques of embryo transfer could be especially useful (see below, chapter 6): a rare antelope, for example, could be induced to super ovulate by treating her with appropriate hormones, and the resulting embryos (fertilised in vitro) might then be implanted in the wombs of some commoner, related species. Only when the second generation is abundant can the conservation breeder begin to relax.

Equalisation

However many founders you begin with, they are of no use at all unless they breed. Indeed, if you want to retain the maximum possible variation within your breeding population, then you should try to ensure that each of your breeding animals produces the same number of descendants. This can clearly be shown mathematically; but intuitively it is obvious.

This is easier said than done, however. Some individuals are naturally far more fecund than others. Some take to captivity better than others. Some find the mates that are chosen for them compatible, and some do not. For these and a dozen other reasons, it is highly likely that some founders will outbreed others; and always possible that one or other of the founders will simply prove sterile. As we will see in chapter 5, both these eventualities beset the captive-bred Arabian oryx. One of the very first founders was sterile. Just a few of the fertile founders have enormously outbred the rest. And so on. One problem was that the Arabian oryx programme began before modern theory was established; theory that clearly shows the need for equality of breeding.

The equalisation of breeding effort must be continued throughout the breeding programme – not just in the first generation. In practice, this means that family sizes should be equalised. Family equalisation means that if one mother has five cubs and another has two, breeding should be from both of the second mother's cubs, but from only two of the first. Again, once the reasons are pointed out, this seems an obvious ploy. Indeed, it can be shown mathematically that equalisation of family size is the most important of all the possible ways of maintaining genetic diversity within a limited population. Again, though, in the early days of zoo breeding it was far from obvious. Indeed it seemed much more sensible (as well as much easier) to breed mainly from the cubs of the more fecund animal. These, after all, were of the lineage that 'did well' in captivity. As we will see, this was the policy pursued in the early years with Przewalski's horse – one of the first species to be bred in captivity specifically for conservation. The most fertile stallions – the 'proven breeders' – were especially favoured. It seemed sensible at the time.

If some animals are too fecund, and if the technology exists, then they can be given contraceptives. If the breeder decides that he really does not want to breed at all from a particular animal – because its genetically similar siblings are already breeding – then he has some hard decisions to make. If he keeps the superfluous animals, then he is using space that could be used for individuals that could supply more genetic variation (or even be given to a different species). One possibility, in some cases, is to try to return such 'superfluous' animals to the wild; though reintroduction should be a highly organised exercise, and the wild should not be used as a dumping ground. Another possibility is to donate the genetically superfluous animal to a zoo that wants to keep the species for educational purposes, but has no breeding ambitions for it. The final option, which often cannot reasonably be avoided, is to cull the superfluous animal.

This last option is of course distasteful, and the only way to countenance it is to cling to the moral generalisation I identified in chapter 1: that it is not always possible to do perfect good; that it is often necessary to settle for the lesser of evils; and that, in this case, it is better to try to save the species by maintaining a viable population than to try to save every individual that happens to be born. After all, a species is not a zoological abstraction. It is a collection of individuals, perhaps many millions, who could continue to wax fruitful for thousands of years to come provided we save the next few generations from extinction.

Several other of the options to suppress superfluous breeding are not so much distasteful as bizarre – at least at first sight. It seems odd to treat a rare animal with contraceptives to stop it breeding; odd to keep a rare animal in a zoo with no intention of breeding it; and positively perverse to kill the offspring of a rare animal in the name of conservation. If such endeavours are to gain public support, the reasons behind them must be carefully explained.

Finally, it is necessary to equalise the genetic contributions of the males and the females. *All* the animals, after all, are sources of alleles. Again, this may be easier said than done. Females exercise choice, and some males are far more desirable than others. Many animals, including many antelope, are naturally polygynous. A single dominant male commandeers many females, while the other males merely look on. To give the lesser lights a chance of breeding, the dominant male has to be removed. But the dominant male is often a very fine animal. His removal creates social disruption. In the wild – and the breeding of animals newly returned to the wild could very well be managed – the lead male may play a key role in herd survival. Here, then, we see yet another possible conflict between conservation (based on sound genetic theory) and welfare.

Ring the changes

Again, even at an intuitive level, it seems obvious that genetic variation is maximised – certainly within any one herd in any one zoo – if fertile animals are allowed to visit other herds. In general (other things being equal), the best possible mix is obtained if every male mates with every female. Males in general should be more mobile than females: males tend to be more mobile in the wild; social herds tend to be stabilised by females, with males as visitors; and it will increasingly become possible to reduce travelling by making use of artificial insemination (see chapter 6).

Sometimes, indeed, the switching of males is desirable simply because particular pairs are just not breeding. It is partly for this reason that Britain's programme to breed Rothschild's mynahs in the late 1980s became so complex, with much to-ing and fro-ing of individuals (chapter 5). Again, though, some directors are reluctant to allow their males to travel, for reasons of welfare. Sometimes they do not want to break up their own social group; sometimes they do not approve of the zoo their animal is travelling to; sometimes they simply feel that the individual animal would be too traumatised. Such considerations must be taken seriously.

Input from the wild

Theory shows, too, that once a breeding population is up and running in captivity, its genetic variety can be increased enormously if just one unrelated individual is introduced from the wild. Of course, endangered animals should not be taken from the wild without careful thought. But on the other hand, many are endangered precisely because the wild has become unsafe for them, and sometimes individuals may actually be rescued from the wild. The present herds of Arabian oryx are still being enriched genetically not from wild animals (for the original wild population went extinct) but from previously unknown herds tucked away in private gardens in the Middle East. New technologies that I will discuss in chapter 6 are also of enormous help. Thus, in principle, and increasingly in practice, wild animals can safely be anaesthetised in the wild, and samples of sperm taken by electro-ejaculation. This can then be introduced fresh into a captive female, or inseminated after freezing, and the wild male is none the worse. Soon, too, it should be possible to rescue the eggs from the entire ovaries of females that have been killed in the wild, for whatever reason.

Generation time

Finally, the oddest breeding recommendation of all – yet one that follows inexorably from the theory. Genes are lost when animals reproduce; there is some loss with each generation. The stated aim is to maintain as much genetic diversity as possible for 200 years. It follows, then, that diversity is maximised for longest if the time between generations is maximised: in other words, if the females are as old as possible when they breed. With this notion in mind, and with the target set – 90 per cent genetic diversity for 200 years – we can refine the ball-park figure of 500: work out exactly how many animals are theoretically needed in the breeding population. The necessary maths is complicated; but when it is done, the conclusions are most intriguing.

It transpires, for example, that 90 per cent of the original genetic diversity could be maintained for 200 years within a mere 37 Caribbean flamingos. They, after all, have a generation-time of 26 years: a mere eight generations over two centuries. Striped grass mice, by contrast, breed at 9 months of age. To get through 200 years requires 270 generations; and to retain 90 per cent diversity through so many generations requires 1,275 individuals at any one time. More examples are shown in Table 1. It follows, incidentally (a point emphasised by Dr William Conway, director of the Bronx Zoo, New York[4]), that it would theoretically cost three times as much to maintain a viable colony of striped grass mice as of Caribbean flamingos. The flexibility of the required number is revealed, too: 500 is far too few if we are breeding mice, but is positively luxurious if we are breeding flamingos. (It seems to me, however, that just about every zoo I know has Caribbean flamingos – which are not endangered; suggesting that zoos do not, in practice, dedicate themselves to conservation breeding as conscientiously as might be hoped.) Note, however, that figures such as 37 (or even 1,275!) do not allow for inevitable injuries and setbacks. Far more are needed for safety.

TABLE 1: Minimum Number Required to Maintain Genetic Diversity

Species	Generation-Time (Years)	Effective Population
Siberian tiger	7	136
Indian Rhinoceros	18	53
Nyala	8	115
Striped grass mouse	0.75	1275
Brush-tailed bettong	6	159
Mauritius pink pigeon	10	95
Arabian oryx	10	95
African black-necked cobra	10	95
Bullfrog	7	136
White-naped crane	26	37
Caribbean flamingo	26	37

Taken from: William Conway 'The Practical Difficulties and Financial Implications of Endangered Species Breeding Programmes' (note 4 above).

One final twist is worthy of note. Zoos that are less than scrupulous, or are not party to modern breeding theory, like to imply or indeed may believe that they are contributing to population simply because their cages are bursting with babies. Yet once a breeding programme is established it is helpful to lengthen the interval between litters; and the animals may serve the cause best that simply stand and wait. On the other hand, zoos that are deliberately retarding the breeding for legitimate genetic reasons should take especial care to explain to visitors what they are up to. It is very easy, in this business, to give the wrong impression!

This, then, is the basic strategy of conservation breeding. The overall point to note is that although zoo breeders, and breeders of prize poodles and Friesian cattle may both call themselves 'breeders', their approach is profoundly different. Indeed, zoos in the past that did not appreciate the differences and bred their animals *as if* they were producing prize poodles did the various species no favours at all. I should discuss this point.

THE PHILOSOPHY OF CONSERVATION BREEDING

Breeders of prize poodles or of domestic cattle have in mind an ideal. *They* decide what the animal they are breeding should be like. In the case of poodles, this is defined – entirely arbitrarily – by panels of judges, who decree that a poodle *ought* to have a forelock of such and such a length, or stand with its legs at such and such an angle, and so forth. Breeders of domestic livestock used at one time to work to similar principles. It has been solemnly written at various times that Berkshire pigs ought to be black with four white socks, and Gloucester Old Spots ought to have just two spots, not one or six or 24. For the past few decades, however, livestock breeders have become increasingly hard-headed, and now define their ideals in terms of milk yield, or growth rate, or fecundity, or what you will; and mere aesthetics has largely gone out of the window.

Whether aesthetics or productivity rules, however, the strategy is the same. Artificial selection is applied. Broadly speaking, the breeder eliminates every individual that does not come close to his ideal. The effect, therefore, is not to retain genetic variation, but deliberately to narrow the gene base. The chief enemy is inbreeding, and in the past many a breeding line collapsed because of this; though many others were rescued by outbreeding before matters became too disastrous. Ideally, though, consciously or unconsciously, breeders seek to select lines that are purged of deleterious alleles, so that inbreeding can be carried on more or less with impunity. They have not of course entirely succeeded, and particular breeds now manifest particular genetic ailments: 'bulldog' calves in Dexter cattle, dislocated hip in many big breeds of dog. In general, though, the aim is to produce near-homozygotes that are more or less free of deleterious alleles yet conform to a Platonic notion of what ought to be. In practice, for commercial or working purposes the pure-bred animals are eventually crossed; for example, greyhounds with collies (or something similar) to produce 'lurchers', and Herefords with Friesians to produce strong beef cattle. But these crossbreeds, though individually heterozygous, are collectively uniform (as with F1 maize).

As we have hinted, zoo breeders in the innocent days of yore worked in this way too. They wanted super-gorillas and super-giraffes to draw the crowds. This was not entirely self-interested. It seemed, after all, that a taller giraffe would indeed be a better giraffe – a 'fitter' animal – than one of more usual size. It is only the more modern field studies that have shown that the animals that do well in the wild may not be the most prepossessing in human eyes: indeed a human just cannot tell what creature will succeed in the wild. Modern genetic theory has shown instead that variety is truly the spice of life.

Yet the desire to hang on to as many alleles as possible itself poses some tricky dilemmas. For instance, Kurt Benirschke and his colleagues at San Diego Zoo at one time had a black-and-white lemur in their breeding population that had a sunken chest. This was clearly a deformity; and it was caused, it seemed, by a single gene. Sunken-chestedness in lemurs is not aesthetically pleasing, and it is hard to see how it might contribute to fitness, but it seemed to do this individual no harm at all. The decision was made, therefore, to continue breeding with this individual, meaning that all future generations will be 'tainted' with the gene.

Yet conservation breeding cannot afford to be too liberal. Some of the few remaining golden lion tamarins (lion tamarins are a group of small American monkeys) harbour a gene that causes hernia of the diaphragm. This condition is potentially weakening, and the aim must be to eliminate this gene while conserving as many as possible of the rest. The present zoo population of Przewalski's horse, now numbering some 700 individuals worldwide, contains several undesirable genes. One of these causes a form of ataxia (uncontrolled movements). This is clearly a 'disorder', reducing fitness, and present breeding plans are designed to reduce or eliminate it. The present captive herd is also 'tainted' with the genes of domestic horses; indeed, a mare was introduced into the breeding stock early in this century to swell numbers. It is also possible that the original wild stock had domestic blood, because the wild horses of Mongolia hob-nobbed to some extent with the domestic animals of the Mongolian peoples. Genes from domestic horses manifest *inter alia* by producing 'foxy' coat-colour (otherwise known as chestnut); a colour never recorded in the wild (though the records are old and incomplete because the wild population has long been extinct). So modern breeding strategy is also designed to eliminate 'foxiness'. This seems to some to be as arbitrary as the pig breeder's desire to produce Berkshires with white feet. But in this imperfect world, the elimination of foxiness seems a reasonable ploy.

In general, too, it is the rare alleles among small, captive populations that should be most officiously guarded. They, after all, are the ones most likely to be lost by genetic drift. Usually the breeder does not know which individuals contain the rare alleles, for most genes do not produce readily visible effects. Mostly, then, the breeder can contrive to save rare alleles simply by trying to ensure that all the founders' genes are carried on. But supposing an allele does produce immediately visible effects? What then? Tigers, for example, carry a gene that reduces the background colour to white; not a gene for albinism, for the stripes remain, but a 'white' tiger. Is San Francisco Zoo, or Delhi, justified in breeding 'white' tigers specifically? It saves a rare allele, to be sure. But the aim (it seems) is like that of the breeders of mega-giraffes. Cheetahs, too, which have precious little genetic variety, do have a gene that merges the spots to produce a blotchy and striped effect. Individuals that are homozygous for this gene, and have this pattern, are called 'king' cheetahs. I know one breeder of cheetahs who would like to maintain a separate colony of 'kings'. Is this justified?

I am not sure I care to provide an answer. On balance probably not, because the enclosure space could in principle be used for something else.

In practice, it probably would not be. And a rare allele is a rare allele. The issue is difficult. That, however, is not the end of the difficulties. For perhaps the biggest problem of all these days in conservation breeding is the most basic of all. Which animals should be conserved?

WHICH SPECIES?

In the fullness of time, if they truly cooperated and committed themselves to captive breeding, zoos worldwide could probably maintain viable populations of *all* of the 2,000 species of land vertebrate that are now thought to be seriously endangered in the wild. So far, however, though there are scores of breeding plans, only a handful of captive populations yet meet the required numbers, and virtually none has an ideal genetic structure; and many animals that ought to be subject to organised breeding, are still out in the cold. The immediate task is to set up breeding plans for the now neglected species as quickly as possible. But it takes time and money and a great deal of organising to set up even a single one (as we will see). So a prime task at present is to establish priorities. Which species should be brought first into the fold?

There are many possible criteria by which to judge. The zoo directors of Europe, who are as good a representative group as any, have emphasised four. First, organised captive breeding is of most use, they say, if the species is endangered in the wild. Given a choice, too, they would rather seek to conserve a species that is unlike any other than one that has a great many similar relatives. Thus the Sumatran rhino would be preferred to the average Amazonian beetle. There are only five species of rhino left in the world, and they are all very different. Thirdly, the European directors prefer to begin serious breeding of animals that are already kept in captivity, but cannot easily be introduced from the wild. Finally, on purely logistical grounds, they would not organise a new breeding plan if it interfered with one that was already up and running.

We might add one or two additional points. For example, some species, sometimes known as 'keystone' species, are ecologically more important than others. Elephants set the tone of entire ecosystems. African tropical forest, for example, is largely of the kind known as 'secondary': the kind that grows up immediately after established trees have been felled. The huge classic trees of the 'primary' forest tend to come along later. But when elephants are around, forest has little chance to reach the primary stage. For this and a dozen other reasons, if elephants disappeared, the entire ecology of Africa would change. The average African mouse (of which there are many species) has no comparable impact. If it came to a showdown, then, we could argue that elephants are more 'important' than any one mouse.

Aesthetics does play a part, too, whether this is morally justifed or not. I have never seen the California condor in the wild, but I have seen many large birds of prey, and they are magnificent. Ideally we would save California condors *and* every Amazonian beetle. But if it came to a straight choice, as for economic and logistic reasons it can, then it would seem perverse to sacrifice the bird for the beetle: like throwing out a Rembrandt to make way for an amateur watercolour.

These, then, are the general criteria by which priorities are decided. But we are left with a conceptual problem. I have spoken airily in chapter 2 of '30 million species' living on Earth. I have suggested that 2,000 'species' of land vertebrates are severely endangered. All this suggests that 'species' is a neat and unequivocal category: each like a parcel tied with string. This, unfortunately, is far from the case. Those who seek to establish breeding priorities must first wrestle with the question, 'What is a species, anyway?'

What is a species?

Plato did not use the word 'species' but he would have had no difficulty deciding what it is. In heaven, he would have said (as indeed he did say) there in an 'ideal' of every kind of thing on Earth – of tables, chairs, and chariots, as well as of living things. The objects and creatures we see on Earth are merely imperfect shadows of their heavenly ideals.

Plato had an enormous influence on science. Deep into the 19th century, biologists believed in their bones, even if they did not state the matter explicitly, that each actual creature that they saw was indeed an 'imperfect' representation of some ideal 'type'; a 'type' that was the model as conceived by God. According to the great American biologist and philosopher of science Ernst Mayr, Darwin had difficulty convincing others of his ideas not simply because he proposed that animals *evolve*, but because he suggested that species could change into other species; and, indeed, that one species could diversify to form many different species.[5] Whether they knew it or not, most of Darwin's biological contemporaries were Platonists, and although they accepted that any one creature, an imperfect shadow of the ideal, could be changed into another imperfect shadow of the same ideal, they could not conceive that one entire ideal could be changed into another. That looked like God changing his mind: an offence against the order of things. But Darwin's chief offence was not against the Church, as is commonly supposed. It was against Plato.

Nowadays, however – following Darwin – biologists feel that there are no such 'ideals'. A species is simply a group of individuals; a very large group, perhaps, but a group none the less. The individuals within the groups between them encompass a huge gene pool, and neo-Darwinians envisage that the gene pool changes as the generations pass, as I have described. The 'species' is as elusive as a candle flame. We can see what it is, and give it a name. But in reality, it is made up of hundreds of individual particles, which are constantly jostling. How, though, do we decide whether two 'particles' belong to the same flame, or to different flames? Or, to be precise, whether two individuals belong to the same species, or different species?

Biologists in general adopt one of two main approaches, depending on whether they are being theoretical or practical. In theoretical vein, biologists tend to define 'species' in reproductive terms. The definition I learnt at university was that 'Two individuals may be considered to belong to the same species if they can breed together sexually to produce fully viable offspring.' At least at a rough and ready level, this works very well. Thus we can immediately see that horses and asses are different species, because when they crossbreed, their offspring, mules, are not fully viable. They are physically strong – they exhibit Darwin's 'hybrid vigour' – but they are sexually sterile, because the chromosomes of the two animals are too different to allow gametes to form successfully. Collies and greyhounds are the same species, however, because although they look very different, there is absolutely nothing wrong with their cross-bred pups.

However, when faced with various individuals whose species they want to decide, biologists are not always able to arrange reproductive tests. In the case, say, of Amazonian beetles, there is just too little time to carry out all the necessary crossbreeding experiments. Palaeontologists seek to establish the affiliations of creatures who are long since dead, and well past breeding. Thus taxonomists (biologists who seek to put names to living things) in practice base their decisions on directly observable physical characteristics: gross anatomical features such as bone shape or colour; the patterns of chromosomes; or the structure of the genes themselves, or of the proteins that they make.

Each kind of approach, the theoretical and the taxonomic, raises conceptual difficulties. Neither quite embraces the true complexities of nature. Neither (alarmingly) do the reproductive proclivities of animals necessarily correspond with their physical characteristics, whether those of gross anatomy or of genetic structure.

Here are a few of the problems.

Subspecies, hybrids, and other complications

Horses and asses and their sterile hybrid offspring provide us with a neat example of creatures that look much the same but are different species:

greyhounds and collies of creatures that look at least as different as horses are from asses, and yet are obviously the same species. These are straightforward examples, however. Life is not always so simple.

Consider, for example, the fire-bellied toad and the yellow-bellied toad of Europe, now being studied by Dr Nick Barton and his colleagues at University College, London. The two toads are very different creatures. As their names suggest, they are different colours. They have different mating calls. They live in different habitats: the fire-bellied prefers large permanent reservoirs, and the yellow-bellied breeds in small, temporary puddles. They are geographically distinct: the fire-bellied lives throughout the lowlands of Eastern Europe, the yellow-bellied in the Carpathian and Balkan mountains. Carl Linneus, the first great modern taxonomist who lived in the 18th century, had no difficulty in assigning them to different species: the fire-bellied he called *Bombina bombina*, and the yellow-bellied *B. variegata*.

But the ranges of the two toads meet, through the Carpathians and around the Danube Basin. Where the two types meet – despite their manifest differences – they mate to form hybrids. Clearly, the hybrids are not quite so 'viable' as either parent species – not for sexual reasons, for they are fertile; but presumably in part because they are not so well adapted either to big reservoirs or to puddles as the parent species. Whatever the reason, Dr Barton and his colleagues have shown that the hybrids do not spread into the fire-bellied's range or into the yellow-bellied's range. Instead, they form a 'hybrid zone' between the two ranges, which is several thousand kilometres in length, but only 5 to 6 kilometres wide.

But how would you classify the two species? By some definitions, they are the same species: they breed together to produce perfectly respectable hybrids. But common sense and observation proclaims that they are different; and the hybrids (apparently!) are not *quite* so viable as either parent species. So biologists in general are content to leave the two as separate species.

Many other pairs of species in nature are known to form hybrids at the meeting place of their ranges – hybrids that fail to spread out of a narrow zone because they cannot compete with either of their parent populations. *Heliconius* butterflies are a classic example, with many different 'races' throughout South America divided by hybrid forms compressed into narrow zones. The hooded crows of North Scotland hybridise with the carrion crows of South Scotland and England, but the hybrids remain in a narrow band through central Scotland.

These are largely academic examples, however, to show the diversity of nature and the dilemmas it can present. The variously bellied toads of Europe and the crows of Britain are not endangered. But the owl monkeys of South America, otherwise known as douroucoulis, present an enormous practical problem for those who seek to breed them in captivity.

Owl monkeys range from Panama down to Argentina – a huge area. As Dr Leobert de Boer recalls in the *EEP Co-ordinators Manual* for 1989,[6] 19th century biologists recognised several different types of owl monkey – differing in size, colour, and distribution of skin glands – but modern biologists have decided that the different types grade into each other, and class them all in a single species, *Aotus trivergatus*. Indeed, it can be difficult to tell the different types apart, so this seems to make perfect sense.

However, when you look closely at the chromosomes, you find considerable variation: from 46 to 56. On the basis of chromosome number, eight different types can be identified. And these differences matter – because hybrids between owl monkeys with different chromosomes are sterile. Furthermore, each of the eight populations with its own particular chromosome number lives in a different place. Since the hybrids *are* sterile; since they also look different (when you look closely); and since the different types live in different places, it is clear that by all reasonable criteria the owl monkeys should be placed not in one, but in eight species.

The theory so far is straightforward. The story is less complicated than that of the fire-bellied and yellow-bellied toads, for example. The problem is that many zoos are breeding owl monkeys, or trying to. Before scientists began looking closely at their chromosomes, and doing proper cross-breeding experiments, it was assumed that they were all much of a muchness. Hence, some present-day owl monkey populations are mixed. This raises several problems. For one thing, success in breeding within those populations will at best be erratic and will be largely deceptive, as many of the offspring will be infertile. Secondly, it may transpire that no one single type of owl monkey is represented in sufficient numbers to form a viable population; so although there may be a lot of captive owl monkeys in total, the group as a whole may be doomed. Ideally, of course, the present captive population would be sorted out into 'pure-breds', and hybrids, and each of the pure types would become the subject of a separate breeding plan. The sorting out would itself be a momentous task. It would also raise an enormous logistic problem; for although zoos may have room for one or two breeding populations of owl monkey, they would not have space for eight. There are other monkeys to consider, after all; and other animals besides monkeys.

The separation of owl monkeys has implications for the wild, too. Suppose, for example, some beneficent Latin American government decided to set up a reserve, which would include owl monkeys. If this was set up in the middle of the range of one particular type of owl monkey, fair enough; the reserve would contain a viable population. But if, by chance, it cut across the boundary between the two owl monkey populations, then neither population might be big enough to be viable.

Monkeys in general are turning out to be an extremely difficult group. As Leobert de Boer points out, Goeldi's marmoset, spider monkeys, capucins, squirrel monkeys and howlers are among those of South America that are now presenting similar difficulties.

Monkeys are probably as variable as they are largely because their forest – though it may all look much of a muchness to the casual observer – is in reality highly discontinuous (at least from a monkey's point of view), so that different populations become isolated in different areas. Each area is then like an island, and as Darwin showed 150 years ago in the Galapagos, each isolated island tends to develop its own suite of species.

Indeed, virtually every group of animals that lives in a discontinuous environment tends to become highly subdivided. Each pond or stream is effectively an 'island' to the fish that live in it – unless the ponds occasionally become joined by flooding. Thus freshwater fish tend to be enormously variable. Sometimes the different variants have drifted so far apart that they can properly be recognised as separate species; sometimes the different populations may look very different, but are still capable of breeding together (if given the chance) and so should probably be placed in the same species.

Thus modern biologists tend to include all the brown trout of Europe within the same species, *Salmo trutta*; but different populations are so variable that since the 18th century (when modern methods of classi-fication began) different biologists at different times have recognised no fewer than 50 different species. Three of the variants – rather beautifully known as the sonaghen, the gillaroo, and the ferox – live side by side in Ireland's Lough Melvin. Dr Andrew Ferguson and his colleagues from the University of Belfast has shown that what keeps them from interbreeding is not their very different looks or their different feeding habits, but simply the fact that each type returns to a different river to breed, and so (by chance) they never get the opportunity to hybridise.[7]

For practical purposes, though, does it matter if different populations that *can* breed together, are in fact allowed to do so in captivity? We have already seen that it can matter: after all, hybrid fire-bellied and yellow-bellied toads, or hybrid crows, clearly cannot compete in the wild with their 'pure' parents. It is true, of course, that animals that mate with others that are too closely related to them are prone to inbreeding depression. But it is also true that outbreeding can be taken too far – leading to 'outbreeding depression'; and enfeebled hybrids, albeit subtly enfeebled hybrids, are a special example of this general phenomenon. We should discuss this further.

Outbreeding depression

We can envisage several possible mechanisms of outbreeding depression. For example, genes operate by producing proteins (or rather, by determining what kind of proteins are produced); and proteins are the body's functionaries. In particular, many proteins function as enzymes, which control body chemistry; indeed there is an old adage in genetics, 'one gene, one enzyme'.

Enzymes, in general, do not work alone. Each one controls one stage in a 'metabolic pathway'. In general, the pathways work most smoothly if each enzyme has evolved in concert with the others; so that each is adapted as precisely as possible to deal with whatever the previous enzyme in the chain produces. Any two animals of the same breeding population will generally have much the same enzymes, and a mixing of genes brought about by crossing will have no ill effects. But if two populations have separated genetically, then they could have enzymes that are significantly different; and any one enzyme may find itself partnered within a metabolic pathway with others that do not behave as it would like. Thus the overall efficiency of the pathway is reduced. This will bring about a small but perhaps significant loss of fitness.

More generally, natural selection tends to lead to adaptation. If a population is divided into two or more sub-populations, then each sub-population is liable to adapt through natural selection to slightly different environments. A hybrid between the two may not be as well adapted to any particular environment, and will lose out relative to the types that conform to one type or the other. This is obvious in the case of the fire-bellied and yellow-bellied toad hybrids. It is less obvious in the case of the carrion and hooded crow hybrids, but no doubt the principle still applies.

Natural selection also shapes behaviour; including the behaviour associated with choosing mates. Clearly, from all we have said, it pays any animal to choose a mate who is different from herself or himself, but not too different. In one experiment, Professor Patrick Bateson of Cambridge University gave quails the opportunity to select mates from between three groups: their own siblings; their cousins; or individuals that were very different genetically from themselves. Given a choice, the quails chose their cousins. We might argue on genetic grounds that first cousins are not the ideal choice – that third or fourth cousins might be safer. But this is the way the experiment was structured – and the point is made; that animals in general will eschew their own siblings as mates (especially if they are brought up with those siblings, and know who they are) but will also eschew individuals who are quite different. Thus, even at an early stage of separation, sheer lack of mutual attraction is likely to impose a reproductive barrier between divided sub-populations. We can see, too, why natural selection should have favoured such a course. It is the obvious way to avoid both inbreeding and outbreeding depression.

We can also see that in animals, once the process of separation has begun (perhaps simply for geographical reasons), then physical differences tend to be exacerbated by behaviour. This follows from what the great British statistician R.A. Fisher called the *runaway principle*. Natural selection (or, rather, what Darwin called *sexual selection*) tends to favour individuals who have visible characteristics that are most attractive to potential mates. Exactly which characters are selected as being 'attractive' is largely arbitrary; with some birds it may be long tails, with others a peculiar way of flying, and with monkeys it may be facial markings. Clearly, though, once these sexual features become recognised in any one population as hallmarks of attractiveness, then the members of that population become increasingly more attracted and attractive to creatures of their own kind who share their views on what is attractive, and less attracted and attractive to individuals in different populations that have evolved different features. That fire-bellied and yellow-bellied toads should mate with one another really is perverse.

Such mechanisms combine to lead separated populations inexorably towards the theoretically total separation that we deem to be speciation. In general, true speciation will of course take longer than any lesser degree of separation. But we cannot in practice discern a time-scale. Sometimes speciation occurs rapidly, and sometimes reproductive union of a kind remains possible even after millions of years of separation. After all, reproductive barriers that may be crucial may be brought about by only one or two genes; but huge genetic and other biological differences may not (by chance) impose any absolute barrier to sexual union.

What is important for conservationists, though, is that the separation is gradual, and at any one time we see pairs or groups of populations at various stages in the separation process: anywhere from complete integration to complete segregation. There is no clear, unequivocal point at which we can say, 'Speciation has occurred *here*.'

In practice, biologists have long recognised the need to give names to the sub-categories that exist within species. Sometimes they talk loosely and colloquially of 'races'. Botanists commonly speak of 'varieties'. Breeders of domestic animals – who impose artificial selection – speak of 'breeds'; and sub-divisions of breeds are 'lines' or 'strains'. But when the differences within wild species are clear to see, and widely acknowledged, biologists tend to divide species into 'subspecies'. Thus among tigers, eight subspecies are recognised; each one having a third 'Latin' name added to the two parts of the Latin species name. Thus tigers in general are of the species *Panthera tigris*. But the subspecies known as 'Indian' or 'Bengal' is *Panthera tigris tigris*, which can be shortened to *P.t.*

tigris; the Sumatran is *P.t. sumatrae*; the Siberian is *P.t. altaica*; and so on.

However, the fact that some species can reasonably be divided into several subspecies raises yet another philosophical dilemma, and another enormous practical difficulty. Throughout this book we have argued that about 2,000 species of land vertebrate (plus a miscellany of fish and invertebrates) could clearly benefit from captive breeding; and we suggested in chapter 2 that if the world's zoos put their weight behind captive breeding, and if societies put their weight behind zoos, then this target should not be out of reach.

But if we now divide those 2,000 species into separate subspecies, and demand that each one of those subspecies should be managed separately, then the number that would need saving could well exceed feasibility. We should discuss this.

What should we do about subspecies?

Why, first of all, might we want to keep subspecies separated in captivity? If they all breed together reasonably well, why not simply re-mix the gene pool?

We have already seen many of the reasons, but it is worth summarising them. First, different subspecies may well be adapted to different environments, and if the aim is eventually to return captive animals to the wild, then those special characteristics should not be squandered. Thus Siberian tigers are bigger, bulkier, and hairier than Sumatran tigers. Hybrids between them might not be well adapted either to Siberian snow or to Sumatran rainforest. Aesthetics also plays a part, whether this is morally justified or not. Siberian tigers are beautiful – huge, robust, shaggy; so are Sumatrans – dark, graceful, slinky. What a shame to squander such qualities in some catch-all hybrid!

But thirdly, and perhaps most seriously, there are as we have seen degrees of reproductive barrier between populations. When the barriers are absolute (or nearly so) we consider their populations are different species. But barriers of lesser degree may still result in hybrids that are less 'fit' than either parent population, and which could well be subfertile. This could prove to be the case with future generations of hybrid orangs, even if there are no obvious problems in the first generation.

Ideally, then, we would want to keep the subspecies apart. So what are the problems? The first is, perhaps, conceptual. As we have already seen – and will emphasise again later – it isn't always easy to decide what is a 'valid' subspecies and what is not. And even when we have made such a decision, we run into a number of logistical problems.

The first and most obvious of these is that there may not be *enough* of any one recognised subspecies to maintain a viable population on its own. The red panda provides a potentially tragic example of this; for as described more fully in Chapter 5, red pandas are clearly divisible into two sub-species, and although there is room for one sub-species in today's zoos, there does not seem to be room for two. By trying to keep both, zoos may in the end keep neither in viable numbers.

With some species, though, captive numbers are so low, and the problems of maintenance are so great, that there is nothing to do but forget subspecies niceties, and mix the few animals there are. Thus, most directors who are interested in the captive breeding of Asian elephants, such as Dr Michael Brambell at Chester Zoo in England, are content to override the distinctions sometimes drawn between the Indian (*Elephas maximus bengalensis*), the Ceylon (*E.m. maximus*), the Sumatran (*E.m. sumatrana*) and the Malaysian (*E.m. hirsutus*). There seems no reason to suppose that there are reproductive barriers between the types, and any ecological distinctions are tiny compared to the appalling demise of all Asian elephants in the wild.

Despite the conceptual problems, then, it is possible to make rational decisions when considering policies for subspecies. Rational decisions do not end the dilemmas, however. Thus, it is only in recent years that zoos realised fully that captive breeding is indeed a serious part of conservation. In the past, their conservational role (if they felt they had one at all) was vaguer: for example, simply to show people at large how wonderful animals are, and thus arouse public concern. Sometimes they took pains to preserve the 'racial purity' of their subspecies – and thus, for example, may have been proud of their reticulated giraffes or their Persian leopards. But in other cases they were pleased if they could get their animals to breed at all, and had no great regard for the niceties.

Thus zoos these days, worldwide, contain huge numbers of hybrids between subspecies. There are dozens of hybrid tigers and leopards; and the fine distinctions that have often been drawn between different lions (Abyssinian, Arabian, and so on) have in many cases been lost. It was recently shown, indeed (by genetic studies) that the world's zoo population of Asian lions was corrupted with African 'blood'.

But what should a serious zoo do with its hybrids? Taronga, in Sydney, is one of several who find itself harbouring hybrid orang utans. They breed well; they are favourites with the public and the keepers. Ideally, Taronga would not want to breed from these animals again, for hybrids are neither Bornean nor Sumatran and they could run into reproductive difficulties in the future. But orangs in zoos can live for 50 years or more. Should Taronga wait for its hybird orangs to die before it begins breeding orangs again – which could take until 2040 AD? With creatures such as fish, it can seem reasonable to get rid of the creatures that are found wanting and start again (as we will see later.) But – is it illogical? – no-one would sweep aside a group of orangs so coolly.

Probably the best solution for zoos that want to breed but who find themselves stuck with subspecific hybrids is to offer them to other zoos that want to keep animals to show to the public, but have no ambitions to breed them (perhaps because they are breeding something else instead). To put the matter the other way around: here, perhaps, is a legitimate role for non-breeding zoos – to act as sanctuaries for animals which, for whatever reason, are not required to breed. The encouraging point is that the hybrid problem should go away. In a few decades' time the present generation of hybrids will have died out, and zoo animals should all belong to definable subspecies – unless, as with Asian elephants, a rational decision has been made to blend them.

All the discussion so far, of the science and the philosophy that flows from it and feeds into it, have been based on a 'classical', Neo-Darwinian view of genetics and populations. That is, genes are regarded simply as abstract entities – 'factors' – that conform to mathematical rules, and yield to common-sense analysis. As we have already seen, this classical approach has got us a very long way. Breeding programmes can indeed be devised to reduce inbreeding depression and maximise genetic diversity; and the considerations of classical genetics do help us to make sense of the peculiarities of fire-bellied toads and of owl monkeys.

Nevertheless, classical Neo-Darwinism with its 'abstract' genetics still requires us to make quite a few assumptions of a kind that should not be taken for granted; it leaves a few loose ends. Some of these loose ends can be tied up by applying the modern techniques of molecular biology and of cytogenetics – which look at genes and the chromosomes that contain them not as abstract 'factors', but as real entities with a distinct and analysable chemical structure.

These techniques are becoming increasingly important; indeed, the fusion of classical genetics and conservation theory with molecular biology has probably been the principal advance of the 1980s; and by the end of the century, every conservation geneticist will be applying the ideas of molecular biology as a matter of course. We should look briefly at these techniques, and at their strengths and shortcomings.

GENES AS MOLECULES

DNA – deoxyribonucleic acid – was discovered in the nuclei of cells in the 19th century by the Swiss chemist von Mischner. (It is not true, as Orson Welles asserted in *The Third Man*, that the only useful thing to come out of Switzerland is the cuckoo clock.) But nobody *knew* that DNA did anything very interesting until the 1940s, when it was definitively shown to be the stuff of which genes are made. Then in the 1950s, as all the world knows, Francis Crick and James Watson, using the data of Rosalind Franklin and Maurice Wilkins, unravelled its structure.

What matters for our purposes is that DNA consists of chains of sub-elements known as nucleotides, each of which contains a base. There are four kinds: cytosine (C), guanine (G), adenine (A), and thymine (T). The sequence of bases is the genetic code. The sequence of amino acids within the proteins that are modelled from DNA – and hence the qualities and capabilities of that protein – is determined by the sequence of bases within the DNA.

We have said, many times, that different animals have different genes. Between different species – say horses and lions – the genetic differences are liable to be great. But between individuals of the same species, too, there are genetic differences. You and I for example are both human beings, with a similar assortment of loci on a similar pattern of chromosomes, yet we undoubtedly have many different alleles upon those loci.

It transpires, however – one of the revelations to emerge since the 1960s – that the DNA of animals is extremely messy. Genes – neat sequences of bases, laying down inexorable codes for proteins – are in practice interrupted by sequences of DNA (known as 'introns') that make no sense as codes at all. They are in fact 'junk' DNA. When the cell makes proteins, it has first to cut out this junk. Between genes, too, there is an enormous quantity of junk. In fact, only a few percent of the total DNA within the nucleus actually functions as genes. By far the majority is junk. (Bacterial DNA, incidentally, is not full of junk. It is good, no-nonsense genes. It is extremely fortunate that molecular biologists worked on bacteria first!)

The junk in animal DNA has been shown to be strange and heterogenous stuff. Among its many peculiarities is the presence at many different sites of what are called 'minisatellites': short sequences of DNA that are repeated, sometimes dozens of times at any one place. The same kind of minisatellite may be repeated many times through the genome; several times on the same chromosome, and often on several different chromosomes.

Yet another revelation is that mitochondria – organelles in the cytoplasm of the cell that are concerned with respiration – also contain some DNA; and that although the total amount of mitochondrial DNA is small, nearly all of it is functional (very little junk), and in fact it helps to create some of the respiratory enzymes.

DNA changes with time; bases get lost, or others are added, or the sequence may be changed here and there. If the changes occur within germ cells (eggs of sperm) they are passed on to the next generation. These changes are the mutations that classical geneticists talk about, and which provide the variation that Darwin perceived was so necessary to natural selection.

If mutations occur within functional genes, then they are liable to affect the anatomy or physiology of the animal. The changes thus caused are liable to be deleterious (a random change in DNA is like kicking a television set) and natural selection will tend to eliminate them. The structure of functional genes, then, tends in general to be highly conserved from generation to generation. But changes in junk do not matter. So mutations in junk are not eliminated. Junk, therefore, tends to be highly variable.

From this description alone we would expect the DNA of different animals to differ in different ways and at different 'levels'; and so we find. Different species – such as horses and lions – have a very different apportionment of functional genes (though we might be equally impressed by the similarities!) as well as of junk. Two animals of the same species will have much the same functional genes; but if they are from different populations within that species they might have significantly different alleles of those genes. That is, some alleles will exist in one population but not the other; and the general frequency of alleles within the two populations will differ. And of course, the junk will differ even more.

Two individuals within the same family – siblings, or parents and offspring – will have very similar alleles, but they will still show differences in their junk. In particular, minisatellite patterns differ from individual to individual so much, that no two people have the same pattern. Hence minisatellite patterns have become the basis of 'genetic fingerprinting', now much used in forensic circles.

Clearly, then, the degree of difference between individuals shows how closely they are related; or at least, if you have three individuals, you can show that two are more closely related to each other than to the third. Clearly, too, because some regions of DNA are so much more conserved than others – and, therefore, some are so much more variable than others – we can use different kinds of comparison to show different degrees of difference. The differences in minisatellites between horses and lions would be so great as to be a total jumble; but if you want to look at such gross differences you can look at genes. The genes of members of the same family are all much of a muchness – but you can spot the fine differences (and the relationships) by comparing minisatellites.

In practice, the DNA of different animals can be compared in various ways. One depends on the fact that single strands of DNA will stick to corresponding strands from the same species – but will stick less well to DNA from another species. So the extent to which a mixture of strands sticks together (which can be measured easily by a change in melting point) provides a rough measure of the degree of relationship. More precisely, DNA samples can be chopped up in different ways into fragments. Different samples break up into different sized fragments – and again, the similarities and differences in the pattern of fragments reflects the degree of relationship. If you really want to, you can analyse the actual sequence of bases within the DNA; but for the purposes of tracking relationships, that is not usually necessary.

Finally, you can, if you choose, explore differences not between the DNA itself, but between different versions of any particular protein, that is produced by different alleles in different animals; or, for big-scale comparisons (for example looking at differences between subspecies) you can look at the chromosomes themselves. Each chromosome (it transpires) consists of a singe 'macromolecule' of DNA, wound round and round and round. In practice, all these methods are employed in modern conservation biology. We should look briefly at a few.

Molecular biology in action

First, at the finest scale, genetic fingerprinting is used in conservation to assess paternity, just as it is in lawsuits. This is valuable for several reasons. First, field biologists have found that many animals are much more promiscuous than is easily observed. Peahens mate with more than one cock in any one season. Hedge-sparrows look monogamous, but in fact indulge in extra-marital sex, and some of their eggs (it transpires) are fathered by cocks other than the one that helps to collect the caterpillars. Male animals that dominate harems do not have life all their own way. Observations on animals as diverse as deer and baboons show that 'sneak mating' is much commoner than the dominant male might like to think.

The trend, now, for reasons of husbandry, is to keep captive animals in the same kind of groups in which they would live in the wild. Extra-marital relations, sneak mating, and general promiscuity make it very difficult under such circumstances to tell who is the father of whom. (We could call this 'the Strindberg dilemma'.) Yet, as we have argued throughout this chapter, we can avoid inbreeding and conserve genetic diversity only if we arrange the matings. In practice, then, genetic fingerprinting (minisatellite analysis) is used more and more, to ensure that particular individuals within the breeding population do indeed have the genetic provenance that they seem to have. Thus, for example, Oliver Ryder and his colleagues from the Zoological Society of San Diego recently described extensive paternity studies among giant Galapagos tortoises in San Diego Zoo.[8]

There are broader issues. Breeding plans depend upon the data recorded in studbooks. But studbooks may be incomplete (zoos in the past did not keep records as well as they do now). More seriously, studbooks records begin either with animals that were already in zoos, or with wild-caught animals. Rarely do they record the precise place that the animals were caught; and the ancestry of the first animals in the book is generally uncertain. In the absence of better information, breeders are obliged to assume that the first animals in the studbook – the founders – are all unrelated. But that in reality may not be the case. The wild-caught animals may all have been siblings. If this is the case, then the best-laid plans to minimise inbreeding may be confounded.

Hence, molecular biologists now employ minisatellite studies to explore the real relationships of animals in breeding groups, to see if the assumptions that have been made about their relationships – based on studbook records – are in fact correct. Thus at this moment, Mike Bruford of the Institute of Zoology in London is employing minisatellite studies to assess the relationships of the Mauritius pink pigeons at the Jersey Wildlife Preservation Trust. The captive population of around 60 birds already shows signs of inbreeding depression. For sure, it has very few founders; the wild population (now being augmented by release of captive-bred birds) was at one time reduced to 106 individuals. It seems likely that at least some of those founders were, in fact, related to each other.

Molecular studies at various 'levels' are also used to differentiate between populations. Conservationists can never save everything; but they try to ensure that whatever they seek to conserve is the most worthy of conservation. Among many animals, it often transpires that there is one 'core' population, and several daughter populations that at some time in the past have split off from the core. In general, the core population will contain the greatest genetic variation, and if there is a choice to be made, then that is the one to save.

The many populations of fish that live in the isolated or semi-isolated streams and ponds in the deserts of the south-western United States provide a nice example. The Dexter National Fish Hatchery in southwest New Mexico at any one time maintains between 10 to 20 species of such fish, and has already restored six of them to their original ranges. Between 1976 and 1985, DNFH raised a great many Sonoran top-minnows; but genetic studies then showed that their particular top-minnows showed no genetic variation – because the captive population had been founded by fish all taken from one (peripheral) population. So DNFH began all over again, with a more varied population.[9]

Broader still, genetic studies of various kinds – augmented by chromosome studies – can help to show differences between subspecies: differences that may not be observable to the naked eye (because animals of different subspecies may look very similar) but which may affect later breeding plans, by producing unviable offspring. Gazelles are proving to be an astonishingly messy group, with several species proving, on closer analysis, to be composed of quite different sub-species. In several cases captive breeding of gazelles has failed, and these previously unsuspected chromosome differences are almost certainly a cause.

Finally, and on the broadest scale, biologists can use molecular studies to assess the relationships between entire species, or indeed families or orders or phyla of animals. Thus, in 1985 a study at the Smithsonian Institute in Washington seemed to solve the ancient riddle of the relationships of pandas. Some say that both the giant and red panda are bears; some say they are racoons; and some say they are neither, and should be given their own family. Stephen O'Brien and his colleagues concluded, from analyses of their DNA, that in fact red pandas and giant pandas are not particularly closely related to each other, and look alike only because they are both carnivores that have become adapted secondarily to eat bamboo. Red pandas (they concluded) are really racoons, and giant pandas are effectively bears – though they diverged from other bears so long ago that they probably do deserve their own family.[10] At this point though we might simply ask, in the words of Juliet, 'What's in a name?'

Molecular studies, then, are becoming extremely useful. Breeding plans in the future must take their findings into account. Yet we should not be carried away. They are not the royal road to wisdom. Taxonomic studies based on molecular studies should not be allowed simply to override all other kinds of information – for they sometimes produce some odd-ball and essentially incredible results. Genetic diversity within a population is one reason for preferring that population, but it is not the *only* conceivable criterion. In short, we still have to use our human judgement.

When all the judgements have been applied, the breeding plans can finally get underway. This raises a whole new set of problems – diplomatic, financial and logistical, which I will briefly discuss.

HOW TO ORGANISE A BREEDING PROGRAMME

From all that has been said so far, it is clear that captive breeding for conservation is bound to be complicated. The underlying strategy, aimed at preserving the maximum genetic variety, is innately complex. Because the population of each animal must be high if it is to be viable, many breeding centres – zoos – must work together, and must often cooperate across national boundaries. In addition to the theory, then (which for most practical breeding purposes already seems good enough), successful breeding needs organisation.

The essence of this organisation is threefold. First, successful conservation breeding depends upon good records: the breeder needs to know exactly which animal is descended from which and is related to whom. The basic data are contained in studbooks – the use of which is abetted increasingly, and necessarily, by computers. Secondly, breeding needs well orchestrated breeding plans; also abetted, in various contexts, by computers. Thirdly, conservation breeding needs organisations of many kinds and at different levels to ensure that the right studbooks are initiated and kept, and that the necessary breeding plans are begun and carried out. Briefly, for the sake of completeness, we should discuss all three.

Studbooks

Dr Peter Olney, besides being curator of birds at London Zoo, is editor of the *International Zoo Yearbook*, alias *IZY* – a job that entails being international coordinator of all the world's international studbooks for wild animals in captivity. He says of studbooks that they are 'the raw material of planned breeding'. And so they are. They contain the essential basic data, in standardised form, on which to base breeding strategies of the kind I have been describing. At present, as recommended in Europe by the EEP (the group that organises breeding in Europe), each studbook should at least list the following information for each individual:

The house name, number, the ISIS-number (which I will explain shortly); the date and place of birth, and the birthweight; the date and place of death, the weight at death and the post-mortem conclusions, and how the carcass was disposed of; who the parents were, and their place of capture if they were wild-caught; the dates of their transfer from wild to zoo or zoo to zoo; and the number and name they have been assigned within their own studbook – which ideally is compounded from their place of birth, plus a number, starting with 1, which denotes when they entered the studbook. These are the minimum data, but studbooks are innately flexible. As time passes, more and more kinds of data might come to be considered 'basic'. The time may well come (though it is quite a long way off!) when a DNA print-out of each individual is considered *de rigueur*.

Of course, long before zoologists began to breed wild animals for conservation, princes, warriors, hunters and farmers bred them for domestic purposes; and the general idea that good breeding depends upon good records (or an incredibly good memory!) goes back a very long way. The first ever studbook – *The General Studbook for Thoroughbred Horses* – was published in 1791. But the first ever studbook for a wild animal was not initiated until 1923, at a meeting convened at Berlin zoo. It was begun by Dr Heinzel Heck on a card index, and was for the European bison or wisent, which had been sadly depleted for some decades and was clearly endangered after World War 1. The European Bison studbook, an historic document, finally appeared in 1932.

Another 25 years passed before the next wild species came on line. This was begun by the Zoological Society of London, at its country home of Whipsnade in Bedfordshire, and was for Père David's deer, and published in 1957. Two years later, 1959, came the third, for Przewalski's horse, the nearest wild relative to the domestic horse. The Przewalski's studbook was established at Prague, where it is still maintained. In 1959, too, a start was made on the okapi studbook, which is now being kept at Antwerp. The okapi, it later transpired, was not as rare in the wild as had been supposed (it is a shy forest creature, which was first made known to science only in 1901) but the other three already depended on captive breeding for survival. Those three, then, the wisent, Père David's deer, and Przewalski's horse, are generally regarded as the 'classics' of modern conservation breeding – although, to be sure, the breeding regimes used in the early days were not always in accord with modern breeding practice.

So the enterprise began to gather momentum. In 1967 the IUCN officially approved the idea of studbooks, and joined forces with the International Union of the Directors of Zoological Gardens (IUDZG) to take responsibility for them. Indeed, these two bodies formed a joint committee which later metamorphosed into the Captive Breeding Specialist Group (CBSG), which is now the grand overseer of all international conservation breeding efforts. The *IZY* of 1967 listed seven studbooks already in train, with another four planned. By 1971 11 were up and running, and another three had been proposed. By 1980 there were 20. *IZY 28*, published in 1989, which covers up to 1986 and is the latest at the time of writing (November 1990), lists studbooks for 95 species: two reptiles, 16 birds, and 77 mammals. Some of these species, however, are subdivided into separate subspecies, each of which has its own international studbook. For example, there are separate studbooks for the Indian (Bengal) tiger; the Siberian; the Chinese; and the Sumatran. In addition to the established studbooks, too, there are a few 'registers'. These, effectively, are studbooks in the making, listing creatures for whom the data are not yet fully gathered.

This is not quite the end of the story. As we will shortly discuss, conservation breeding for practical purposes is for most species carried out on a regional basis, rather than a global one. Thus, the various regions each keep studbooks of at least some species that are not contained in the *IZY* international list. For example, studbooks for many Australian species are at present confined to Australia. Taken all in all, however, we are still a long way short even of listing the basic data that would enable us to breed in captivity the 2,000 or so species of land vertebrate (or 2,500 or so subspecies) that could theoretically benefit; and of the creatures that are properly registered, only a few (such as Przewalski's, Père David's deer, and the Siberian tiger) are yet kept in numbers that could be considered comfortable. The pessimists already say 'It's hopeless!' But the optimists point instead to the exponential rate of the growth of studbooks, and anticipate that by the beginning of the 21st century we will have several hundreds, which is a reasonable proportion of what is needed, and a great deal better than nothing. In this business, optimism is necessary; and things are not quite so hopeless as to make it entirely inappropriate.

Each individual studbook rapidly grows in complexity. It contains, after all, the raw data of a family tree (or, rather, of several family trees) and anyone who has ever drawn up a wedding list knows how rapidly such trees begin to addle the wits. The computer, therefore, has been a godsend. Indeed (as with modern insurance and various forms of tax) it is hard to see how conservation breeding could have moved into its modern phase without the computer. The first truly significant computerising step was taken in 1971 (early days in modern computer technology!) when a group of enthusiasts in the United States set up ISIS: an acronym that orignally stood for 'International Species Inventory System', but has now been changed to 'International Species Information System'. ISIS is centred now at Minnesota Zoo in Minneapolis. The central role of ISIS was not, in fact, to produce studbooks, but simply to keep an up-to-date and accessible central list of all the animals in all the zoos that chose to make use of its services. In theory, such information is useful to directors even if they are not compiling studbooks or taking part in organised breeding plans – for zoos sought and seek to breed animals even in the absence of an overall plan, and need to know who else is in the field.

In practice, the ISIS data cannot by itself provide all the information needed to establish a truly comprehensive studbook. Its basic data are provided by subscriber zoos, and there are many genetically valuable animals in zoos that are not subscribers. But once the studbook is established (the non- subscribers having been enquired into) the data from it are added to ISIS. So ISIS is the best international centralised general data pool of zoo animals that the world yet possesses; and if there is ever to be a truly comprehensive world inventory, then this will be it. Now, too, there are many computer programs, with many acronyms, in many countries, containing studbooks for individual species, and enabling the lists of species to be analysed in various ways (for example to calculate inbreeding coefficients). But ISIS is the daddy; and many other programs pick up on data on their way into ISIS, or make use of what comes out.

In general, computers and their software develop so fast that it hardly seems worth even trying to discuss in detail what is going on. Suffice to say only that if you can think of a way in which computers might in theory help the business of keeping and anlysing data relevant to organised breeding, then somebody somewhere probably has or is working on a program for it. The main obstacle at present seems to me to be that of computer compatibility, exacerbated by the rapid development of hardware and software. At present, some of the zoo world's computer experts spend much of their time translating programs between systems. But this will settle down.

I have mentioned a few of the organisations that now control international breeding plans. Again, exhaustive detail seems superfluous but a brief survey of the main organisations, and the breeding plans that emerge from them, is appropriate.

Organisations and breeding plans

The organisation of animal conservation worldwide seems at first sight horrendously complex; a bureaucratic nightmare; an explosion of acronyms. Perhaps it would benefit from some tidying up, but on the whole, complexity seems inevitable. It makes sense to organise conservation at various levels: international, regional, national, local. It makes sense for different groups to pursue different ends; the World Wide Fund for Nature (WWF) concentrating on funding, the IUCN on organisation. Then there are various *ad hoc* groups with a pioneering past and a very active present. One of these is the International Council for Bird Presentation/International Waterfowl and Wetland Research Bureau (ICPB/IWRB) – which, *inter alia*, as we will see in the next chapter, is now coordinating the reintroduction of the Bali starling, alias Rothschild's mynah. Another is the Fauna and Flora Preservation Society of Britain (FFPS), which initiated the salvation of the Arabian oryx, forever the exemplar of captive breeding. There is no call for such influential groups to disassociate, just in the interests of bureaucratic tidiness. It makes sense, too, for different lobbies to group together, such as zoo directors, nationally and internationally; and it makes sense to appoint committees to attend to particular species or groups of species, each of which acquires its own acronym. It is inevitable that particular individuals should sit on more than one committee, and sometimes apparently on dozens. Without being Panglossian, then, and arguing that all is for the best in this, the best of all possible worlds, we may concede that complexity of organisation is unavoidable; and learn to live with the proliferation of letterheads.

In general, though, the organisation of most relevance to this book is the IUCN, a subgroup of which is the Species Survival Commission (SSC), to which the Captive Breeding Specialist Group is answerable. The CBSG itself is, as we have seen, compounded both from IUCN and the International Union of the Directors of Zoological Gardens (IUDZG), and is chaired by Dr Ulysses Seal – who is, in fact, a medical doctor at the Veterans Administration Medical Centre in Minneapolis, and is also prodigiously energetic. The CBSG, based in Minnesota, is truly the focus of the world's conservation breeding efforts.

As Peter Olney said, the breeding plans themselves rely upon the studbooks for their raw data; but a list of data by itself does not a breeding plan make. The plans have to be made separately. For practical purposes, they are for most purposes conducted on a regional rather than global basis – the point being that animals should not have to travel between zoos further than is absolutely necessary, and in many cases, travel is subject to several kinds of legal restrictions, including veterinary. This means that for practical purposes, each region keeps its own studbooks, each of which of course contains part of the data in the corresponding global studbooks. Only for a few creatures, such as the giant panda, is it truly necessary to breed on a global scale, simply because there are so few giant pandas outside China. However, we could legitimately wonder whether it is justified to keep pandas outside China at all, as they breed so badly in most of the world. Possible political instability in China seems to be the main legitimate justification for trying to breed elsewhere.

The regional breeding programmes, which are by far the majority, are as follows: North America has its Species Survival Plans, or SSPs, administered centrally by the American Association of Zoological Parks and Aquariums, known as AAZPA, and based in Bethesda, Maryland. In March 1989 there were 50 SSPs in progress. Europe has EEPs, which can be translated as European Endangered Species Breeding Programmes, but which properly stands for *Europäische Erhaltungszucht Programme*. There is no central body comparable to AAZPA to whom the EEPs are answerable, but there is a coordinating committee. EEPs involve East European as well as West European zoos, and have already spread beyond Europe, into Israel and Africa.

The British Isles have plans administered by the Joint Management of Species Group, answerable to the National Federation of Zoos of Great Britain and Ireland. Its List A species are already the subject of active breeding plans, while its List B species are monitored, and could become List A species. Many people feel that the British Joint Management schemes should and will become part of EEPs; and indeed, some EEPs are already coordinated from Britain. The present separateness is not another example of British insularity. The British plans were established before the EEPs.

Australia and New Zealand has its Australasian Species Management Scheme (ASMS), administered by The Association of Zoo Directors of Australia and New Zealand, or AZDANZ. Japan and India, too, each of which is a region, have already taken the first steps towards conservation breeding, by establishing studbook keepers. The Japanese Association of Zoological Gardens and Aquariums (JAZGA) did so in 1988 and the Indian Zoo Directors Conference did the same in 1989. In the fullness of time, the land masses of Europe, Asia, and Africa should become more and more closely coordinated. Studbooks, then, abetted by computers, provide the material. Various administrations provide the coordination, with the CBSG at the centre. How, in practice, does a breeding plan get off the ground? EEPs can serve as the model.

How to begin an EEP

EEPs began late; they were initiated only in May 1986 at a meeting in Nuremberg of continental members of the IUDZG and the German Association of Zoo Directors. But now (1990) more than 30 EEPs are in train and Europe should soon rival North America in its organised contributions to captive breeding. The way EEPs should be organised is described by Dr Christian Schmidt in the *EEP Coordinators' Manual*. Item one is to decide which species or subspecies to take on board. This is decided by meetings of European zoo directors, according to the criteria we have already discussed. Item two is to elect a Coordinator for the Species. The Coordinator could well be the person who is already keeping the international or regional studbook for the species, though it may not be; and indeed, a species may be nominated for an EEP even if there is, at the time, no studbook in existence. The Coordinator has to be confirmed by all the member zoos (that is, the zoos who keep the relevant animal) and stands for re-election every three years.

In addition to the Species Coordinator and the studbook keeper (who may be the same person) there is a Species Commission. This of course includes the Species Coordinator (who is, in fact, its chairman) and the studbook keeper, plus three to 12 elected representatives from the zoos who hold the particular animal. Every zoo in the EEP has to have a species representative, whether or not that person actually sits on the species commission at any one time. Election to the commission is for three years (which is renewable). The Species Coordinator also makes and maintains contact with the coordinators of any equivalent SSP or JMSG plan, and with the chairman of the SSC (in the IUCN) that is concerned with the particular species. Only by such formalities is it possible to maintain the necessary worldwide communication and cooperation.

If there is no international studbook when an EEP is to be launched, then the Coordinator applies to the International Studbook Coordinator (that is, Peter Olney at London Zoo) for permission to begin one; which he grants on behalf of IUCN and the IUDZG. But even if there is an international studbook, the Coordinator still has to ensure that a regional studbook is set up, containing data already outlined above. In the course of setting up this bureaucratic framework, the Coordinator will have written to all zoos in the region (or in feasible interchange distance) that hold the particular animal, to ask if they would care to participate in the EEP, and what the zoo can contribute. Zoos that accept agree to play by EEP rules for that species for at least six years. They must also of course give full details of which animals they have, how many they are prepared to keep, and what accommodation they can offer.

The Coordinator's next letter to the holding zoos reveals which ones are actually taking part, which animals they have, and so on. It is at this point that the species commission is set up. But then, the Coordinator and the commission have to explore the matter of subspecies; and will generally decide to focus on just one or two subspecies. As we have already discussed, however, this decision can have many ramifications.

With the animals identified, the studbook underway, and the zoos ready and willing to take part, the next step is to access the animals themselves: their genes, their demography, and their state of health. On the genetic front, the coordinator must identify the founders (whether living or dead), and then calculate the percentage of each founder's genes in the offspring that those living animals would produce in all the different possible combinations (a pretty hopeless task without a computer!). He or she then has to determine the number and sex of the animal now breeding, and the number of offspring of each living animal. From all this information he calculates the inbreeding coefficients of living animals and of the total population, and the inbreeding coefficients of the hypothetical offspring that would be produced by all possible combinations of the living animals. Finally, he should undertake a direct genetic analysis of each animal, by analysis of DNA and by karyotyping. The relevance of all this is evident from all that we have said so far in this chapter.

Demographic considerations are equally important. The first and obvious requirement is to determine how many individuals there actually are in captivity, in how many zoos; how many enclosures might be made available in the future; and how many similar creatures would be ousted from similar enclosures if the population expanded as required. The coordinator then works out the optimum population size – which in effect is a compromise between the 'carrying–keeping capacity' of the zoos (how many they can look after between them) and the number that theory demands: a number that will be in the region of 500, but is modified, as we have seen, by consideration of generation time. The age and sex of each animal is also important; and the age-specific rate of survival and fertility (that is, how long an individual of a given age is liable to live and how many offspring it is liable to produce); and from these, the coordinator can judge when the optimum population size is likely to be reached. Finally, when the optimum population size is reached, decisions must be made on which animals to keep in the breeding programme, and which to treat with contraceptives.

Overall, the strategy of an EEP runs as follows – again in accord with all the theory we have outlined so far. The population must be brought to its optimum size as rapidly as possible, and then be kept stable. The percentage contribution of the founders must be equalised (equalising family size); the males and females equalised if possible; and matings designed to avoid high inbreeding coefficients. Animals should be exchanged for breeding only with other zoos in the EEP or another organised breeding programme, but as far as possible without too many quarantine boundaries. Pains should be taken to exclude genes that are obviously undesirable – a matter which, we will see, is of great relevance with Przewalski's horse. Animals should be kept in their natural social groups – those that live in harems kept in harems. Zoos should be favoured that are experienced, have good enclosures, and are prepared to exchange animals without buying or selling. Zoos have to agree to cooperate, of course, but they are entitled to object to a proposed transfer. If they do, though, they can be overruled by simple majority of the species commission. On the other hand, any particular transfers that member zoos suggest should be followed if this is compatible with the overall plan.

Finally, individual animals in an EEP must be easily identifiable at a distance, which is liable to involve some system of marking; and the commission is also expected to produce guidelines for husbandry. In short, it is not easy to organise a proper breeding programme; and the coordinators in particular (who generally have full-time jobs of their own, within a zoo) are kept busy.

This, then, is the underlying theory of conservation breeding, and an outline of the way that breeding is organised. In the next chapter we will see what is actually happening.

FIVE

PROJECTS IN PROGRESS

It would be nice to write a tidy chapter of achievements, with a neat and inexorable prospectus of future projects. It would be good to feel that the 2,000 or so land vertebrates that would clearly benefit from organised captive breeding will all receive the help they need before it is too late for them – plus a scattering at least of fish and invertebrates, even though these animals cannot be so comprehensively accommodated. It would be nice to anticipate, too, that all the captive creatures that need to be returned to the wild will be steadily restored as the world becomes kinder.

It would be nice: but life is rarely is simple as this, and in the world of captive breeding, it certainly is not. I hope I have said enough in previous chapters to show that the complexities are immense; not only the science (which is straightforward enough for practical purposes – at least when computer-assisted) but also the organisation. A huge number of separate items need to be attended to. Each species (or subspecies) needs a studbook keeper and a species coordinator. Each breeding plan must involve at least several zoos, and often a great many, in different countries. Each country should coordinate its efforts with every other. Rich countries must negotiate with poor, even though both may have other things on their minds (the Gulf War has been waged while I have been writing this – and indeed has altered various conservation plans). Those who understand and advocate the kind of approaches to conservation outlined in this book must argue with those conservationists who mistrust all science and apparently find melancholic solace in the decay that they perceive to be inevitable; the California condor, some said, should be allowed to 'die with dignity'.[1] Every decision of lasting value has in practice to be made by a committee. Progress is through international workshops, which all have to be convened. The Anglo-French Channel Tunnel is not more complex in its number of parts than this; but the Channel Tunnel has a budget of billions, and the prestige of two governments behind it, while the conservation of wild species by captive breeding is, for most of the world's economies and their governments, a marginal activity.

But the problems of science, organisation, and finance are simply those that are intrinsic to the enterprise. Even in an ideal world, with all the decks cleared and everyone pulling on the same rope, those problems would still be enormous. At present, however, the decks are far from clear. Many zoos worldwide are still run by entrepreneurs or – the bane of many a municipal zoo – by local politicians on the way up or on the way down, who really do not appreciate the problems, or the urgency.[2] Even good zoos still encounter enormous local and political opposition, largely (I believe) because they often fail to tell people what they are really doing and why, and still act as if they were run by entrepreneurs.

There is also the pressing practical matter of enclosure space: very few captive populations are liable to reach the hundreds required for statistical safety unless every enclosure in every zoo worldwide is earmarked. But some zoos do not want to be involved; some feel (usually mistakenly, I feel) that they have to keep the standard livestock – one elephant, one giraffe, one lion – to keep the visitors coming in, who are a *sine qua non*; and many – even the best – are reluctant to get rid of favourite, thriving animals simply because those animals are not required for coordinated breeding purposes. Many zoo animals are hybrids, usually between subspecies: Bornean x Sumatran orangutans; Sumatran x Siberian tigers. Monkeys (which have a marked tendency to form subspecies within their native forests), and gazelles (for some reason) are a particular mess. Some other zoo animals are almost certainly pure-bred, but their pedigrees were lost in the days when organisation was not so tight as it is now becoming. Species coordinators generally feel that all such animals should be left out of the breeding plans, but they continue to occupy space.

So life in reality is difficult. There are many successes; plenty of scope for encouragement; but so far, there is no tidy and numerically convincing list of missions accomplished. There is even a small paradox: that we cannot honestly register success until a particular project has been up and running for a few decades, and we can see how things have turned out; but the projects that began before the late 1970s (or, some would say, before the late 1980s) generally contained conceptual flaws because they began in the days before the necessary theory and technique were properly worked out. So within some of the best established and rightly admired schemes (such as those for the Arabian oryx and Przewalski's horse) there is now some element of rescue, to correct genetic trends that in an ideal world should not have been allowed to arise. The secret, however, is not to give up. Journeys of a thousand miles begin with a single step, as Mao Tse Tung observed. Hope is one of the necessary virtues that St Paul identified.[3]

But in an attempt at tidiness, this chapter will take an overview. It will identify the logical sequence of events in the creation of a convincing captive breeding scheme and illustrate each stage with a few conspicuous examples. So what, first of all, is the logical sequence?

Stage one is to decide what species, or subspecies, to make the subject of captive breeding, and then to appoint a coordinator and studbook keeper: Once it is accepted that such and such an animal might benefit from captive breeding (if this is not already happening), then the next stage is Population Viability Analysis, or PVA. Biologists look at the animal in the wild, to try to assess how well it is doing, and what its prospects are for doing better.

Out of the PVA comes the survival plan, which is most likely to be hatched at a workshop convened by Ulie Seal of the Captive Breeding Specialist Group of the IUCN. The best experts who can be brought together decide what should really be done – bearing in mind that knowledge can never be comprehensive; that we will always be obliged to rely, as a physician must do, on the best informed opinion. Might it be best simply to leave the animal in the wild? If captive breeding is called for, then how many individuals should be brought in, and how many left in the wild? Where should the captive animals be placed? In reserves in their own country, in zoos or reserves abroad, or both? Who has the space and the wherewithal to look after them? These decisions are not easy, and, when made, will never be unopposed; as we will see later from several examples.

Then, the animals that are taken from the wild, together with any that are already in captivity, must be marshalled into a convincing breeding herd – usually on a regional basis, but occasionally (*vide* the giant panda) on a world basis. This is done by the species coordinator assisted by (or doubling as) the studbook keeper. Then the breeding proper begins, starting with founders who are considered to be most worthy: those that are not hybrids, which are not too closely related to each other, and so on. The first requirement is to raise numbers to a hypothetically 'safe' number, while maintaining the widest possible genetic diversity; and this must be done as quickly as possible, to reduce genetic loss by genetic drift, as was outlined in chapter 4.

If and when the breeding programme starts to go well, demographic problems begin. Of course, to conserve genes and ride out natural hazards, populations must be large; but they must also be accommodated within the spaces available to them. A population of 300 animals that is poised to double every two years is an embarrassment, if only 350 spaces are available in the region. To conserve genetic diversity in a population of indefinite size is not easy. To conserve it within a population that is deliberately kept within (usually fairly modest) bounds requires tight management and cooperation. Sometimes the problems of numbers seem to arise almost as soon as those of reproduction are solved. Such was the case with the golden lion tamarin, as we will see. But within virtually all successful breeding programmes, demography rapidly becomes a pressing concern.

When the captive population is at least self-sustaining, reintroduction becomes an option. To be sure, this option will not necessarily be acted upon; or at least not immediately. Successful reintroduction is far from simple, and there are many obstacles to overcome before it can begin. There is also a logical constraint. Ideally, captive breeding should begin before, and preferably well before, the wild population is beyond redemption. Neither should captive breeding be seen as a substitute for habitat protection or for management in the wild. Thus, in general, captive breeding and protection in the wild should proceed side by side. Ideally, both programmes would succeed, in which case the option of reintroduction would not need to be exercised, although individual animals might be exchanged between the wild and captive populations to ensure that each is genetically enriched. It is an inescapable and oft-demonstrated fact, however, that protection in the wild does not always succeed – and probably, for many species, it cannot succeed in the short term; and of course, if it does not succeed in the short term, then there will be no long term. In many regions, the possibility of war alone is enough to justify captive breeding programmes elsewhere.

Even if reintroduction is deemed necessary and desirable, then in many cases – most – this phase is held up, because there is nowhere safe or suitable to return the animals to. In many cases we must simply wait until – as suggested in chapter 2 – the human population falls (one hopes of its own volition) and wilderness again becomes available. But sometimes, even now, space can be made available. Sometimes all that is required is a change of heart among the local people or government (sometimes governments lead, and sometimes the people lead). So it is that more than 100 introduction schemes are already in progress worldwide – which between them raise and illustrate a whole spectrum of new problems.

Finally, the animals, once returned, must constantly be monitored; to ensure that they are fit enough as individuals – perhaps after several generations as captive animals – to cope with the wild; to ensure that the habitat that has been prepared for them is truly suitable – for it will not always be the one they originally lived in, and in any case will generally have changed since they left it; and then to ensure that the wild herds (which will usually be smaller and more confined than the original wild herds) do not run into further genetic problems. The reintroduced herd, in short, will inevitably be managed to a greater or lesser extent; which indeed must be the case for most large animals (and many small ones) if we truly value their survival. Increasingly, there will be a flow of genes (either individuals or simply of gametes – see chapter 6) between the wild and captive populations.

All this, at least, is the ideal. It is hard to find a single convincing example in which the entire logical sequence has been worked through from beginning to end – mainly because these are still early days. To be sure, several species have been rescued from the wild, bred in captivity, and then returned, including the European bison, Père David's deer of China, the Arabian oryx, the Hawaiian goose, the red wolf of North America, the gold lion tamarin of Eastern Brazil, and several more. Several more have been bred successfully in captivity and are now on the point of return: Przewalski's horse of Central Asia, the black-footed ferret of North America, and perhaps the California condor are among them.

But the programmes of capture and reintroduction that are already well advanced are obviously the ones that began some years ago; and none of those mature examples began under ideal circumstances. There was no PVA to decide whether a captive breeding programme was really required, and how this would leave the remaining animals, and which animals should be taken out. The classic examples were not cool exercises in applied conservation science. They were last-ditch rescues. The last few Arabian oryx were about to be shot when they were captured; the last few California condor were being poisoned; and the last few black-footed ferret were threatened with canine distemper. All of them were simply snatched from the jaws of oblivion. Furthermore, some reintroductions (for example of the Hawaiian goose) cannot be considered an unequivocal success. On the other hand, those animals that have been subject to modern PVA and all that follows – such as the Javan rhinoceros and the Florida panther – are not yet involved in up-and-running breeding programmes. But at least now, as the 1990s get under way, we know (or think we know) what ought to be done.

I will look at the various stages of an ideal captive breeding plan in turn, to see how things are progressing; and to see whether and to what extent reality can ever match up to our ideals.

Decision time for the Javan rhino

A contentious issue at the beginning of 1991 is what to do about the Javan rhino. This is one of the three species of rhinos in Asia. Like the Indian, but unlike the Sumatran and the two African species (the white and black), it has only one horn. It is probably the rarest species of large animal in the world. There are probably 50 left in Java (no one is quite sure), all in the tiny peninsula of Ujung Kulon to the west of the island; and perhaps ten in Vietnam. There are none in captivity, in Java or anywhere else.

In general, Indonesian conservationists such as Widodo Ramono, and Charles Santiapillai of WWF who is based in Indonesia, would like conservation efforts to focus on Ujung Kulon itself, primarily to protect against poachers.[4] A second population should be established, they say, but within Indonesia. The population has appareently been stable for the past 15 years, they say. The composition of the present population (how many males, how many females, how many of breeding age, how many in breeding condition) is unknown; and so, too, is the actual and potential breeding rate. Any animals that were caught for captive breeding would have to be caught randomly – they are difficult enough to see, let alone catch – and those that were left might well be a non-sustainable rag, tag and bobtail. Among captive rhinos in general only blacks are now breeding reasonably reliably. In particular, the attempted rescue of the Sumatran rhino has not so far set an encouraging precedent. Sumatran rhinos in the wild are now reduced to around 600 animals – not so parlous as the Javan but none the less already split into sub-populations that in general are not individually viable. Small captive herds of Sumatrans have now been established in Sumatra, and divided among several zoos in America (including San Diego and Cincinnati) and Port Lympne in England; but none have bred, and several have died during or after capture. No Javan has ever bred in captivity, and at present there are none in captivity. Capture, transport and captive breeding are horrendously expensive (each Sumatran in captivity has cost hundreds of thousands of dollars) and, say Drs Ramono and Santiapillai, the money would be better spent on local protection than on captive breeding schemes that they perceive to be highly speculative. On the face of it, the conservative school seem to have a very strong argument.

Ulie Seal of CBSG concedes that these arguments are indeed strong; that all the dangers of capture and shortcomings of captive breeding must be taken into account. Nevertheless, he feels that future strategies for the Javan rhino must include captive breeding. In practice he recommends that between 18 and 26 animals should be removed from Ujung Kulon, and these should found two more captive herds in Indonesia, and one in some remote area, safe from any war or natural disaster that might overtake Indonesia itself – which in that part of the world could include volcanoes.

This recommendation rests upon several lines of thought. First, there is the general notion that emerges from the theory in chapter 4: that a population of 60 is more or less bound to go extinct within a few decades, if it is simply left to itself. If it does not fall foul of natural disaster in the short term, then it will become increasingly inbred. In small populations that remain small for several generations the rate of genetic drift is enormously high. Ulie Seal advocates, indeed, that the population of Javan rhinos should be raised as rapidly as possible to 2000 animals. Even without the insight from detailed ecological studies, however, it seems highly unlikely that the Ujung Kulon population could ever rise above 100; and perhaps the present 50 or so is near the upper limit. It is tempting to calculate the area that an exotic species needs by drawing comparisons with familiar animals in familiar circumstances. Thus we may observe that a domestic cow can get by on about half a hectare, even when it is lactating and its energy needs are high. Javan rhinos are roughly five times as heavy as a cow – so we might casually suggest that one rhino could get by on about two and a half hectares. But cows have evolved to eat grass, whereas Javan rhinos are browsers. The grass that cows eat on modern farms has been bred to be nutritious and palatable, and every non-edible species is rigorously excluded from the pasture. Perhaps only one in 100 of the plants that surround Javan rhinos are edible. *All* those plants in general do their best to avoid being eaten, and are liable to be tough and probably toxic. Only a small part of each plant can be eaten. So if a cow needs half a hectare of lush pasture, we might conservatively estimate that a rhino in a wild forest would need five times more (to take account of size difference); another 100 times more (to take account of the proportion of plants that are edible); and 100 times more again (to take account of the proportion of each edible plant that can actually be eaten at any one time). So we have two-and-a-half times a hundred times a hundred = 250,000 hectares of forest per animal. The real figure is bound to be much less than this, for browsing animals know their own business better than I do. But the principle holds: the area each animal requires is liable to be vast.

The rhino population of Ujung Kulon has apparently been steady for the past 15 years. This might be a good thing – or not. Rhinos are not rapid breeders by some animals' standards, but if they are protected and conditions are good, then their populations can and do expand rapidly. Thus Kenya now has roughly 300 rhinos among 11 dedicated reserves. Rob Brett of the Kenya Wildlife Service and the Zoological Society of London reports that between 1986 and 1989 black rhino populations within those reserves have grown by as much as 15 per cent per year; while the slowest rate of increase was 3 per cent. Over the whole 11 reserves, the average increase has been around 10 per cent per year. Dr Brett estimates that Kenya's dedicated reserves could hold about 600 animals, and that this number – which is beginning to achieve comfortable dimensions – should easily be achieved in ten years.[5]

So why isn't the Ujung Kulon Javan rhino population increasing?

Partly because it is being poached (slowly; but a few losses make a big difference to such a population); and partly – indeed certainly – because the area simply is not large enough to accommodate more. Thus, if any new rhino is born (and we do not know how many have been born) then, sooner or later, that newcomer or some other rhino will be pushed out; which means it will die.

All these factors – the likelihood of disaster, such as an erupting volcano; the breeding potential of the remaining animals (in so far as this can be judged); the holding capacity of the environment; the rate of genetic loss by genetic drift and the effect of this (if known) upon subsequent breeding success – are quantified as far as is possible, and all these quantities are fed into a computer to become part of the Population Viability Analysis. If all the relevant factors could be quantified exactly, then the PVA would give precise bookmaker's odds on the population's chances of survival over a given period. In practice many of the important parameters can only be guessed (albeit sensibly), so PVAs give only a range of likelihood – which in the case of the Javan rhino is quite a wide range. None the less, the conclusion from PVA is that the Ujung Kulon population left to itself is doomed. It cannot expand within its present confines; but unless it expands, extinction will inexorably ensue.

There is one last common-sense argument to suggest that some animals must be removed from Ujung Kulon: one that underpins the world's most sensibly controlled fisheries. If a population is as big as its habitat will allow – if it is at 'carrying capacity' – then, by definition, it cannot expand. So fishermen who wish to take the 'maximum sustainable yield' – which again by definition is the most that can be produced over an open-ended period of time – must first *reduce* the parent population. If they do not, then there is no room for expansion; there will be no 'yield' for them to take. If they first reduced the parent population by only about 10 per cent, then they would not get a large sustainable yield, because there would not be much room for expansion. If on the other hand they first took out 90 per cent of the parent population, then they would not get the greatest sustainable yield either, because the parent population would then be too small to produce many offspring. It transpires, indeed, that the maximum yield is produced by first removing about half the original population; the natural compromise between too few and too many.

I take this line of thinking as independent evidence that Ulie Seal's PVA of the Javan rhino is at least sensible. The aim is to produce 2,000 animals as quickly as possible. The Ujung Kulon population cannot expand unless some animals are removed. If about half of them are removed (Seal's 18–26) then this will allow the remaining animals to expand at the maximum rate possible. If the animals that are taken out also breed, as they should, then that will be a bonus. Even if they did not, however, then – if fisheries are any precedent – the population in Ujung Kulon could be the same in ten years' time as it would have been if none had been taken out. By this argument, there might be nothing to lose by removing some of the animals.

What of the lack of success so far in the Sumatran captive breeding programme? Is this not a discouraging precedent? Certainly it is; but there is no reason to suppose the present problems cannot be overcome. Sumatran rhinos have been caught in the wild so far by digging pits, and some have been injured along the way, which is a terrible thing to do to an endangered animal (or any animal). In Africa, both black and white rhinos are commonly caught and translocated from place to place, and they are usually caught by anaesthetic darts. In fact, although this whole process nowadays generally works wonderfully, anaesthesia cannot be taken for granted, particularly in the field; for example, the effect of an anaesthetic depends very much on the animal's state of mind. We cannot assume, therefore, that 'chemical restraint' is necessarily preferable to mechanical constraint, and if mechanical constraint can be made to work well (holding the animal 'gently but firmly') then it is probably preferable. In short, it is not unreasonable to catch rhinos in pits if they have to be caught at all; but we should be very careful how we design those pits. Method of capture, in general, is still an issue.

It is true, too, that Sumatran rhinos have not bred well in captivity, and indeed that among rhinos, only the black seems to be breeding satisfactorily in zoos. But it is hard to conceive that this state of affairs will persist. As we will see in the next chapter, scientists at London's Institute of Zoology are now providing the basic physiological insights that should make it possible to encourage breeding with far more certainty; and Cincinnati's Center for the Reproduction of Endangered Wildlife (CREW) plans to apply the advanced techniques of artificial insemination and embryo transfer. CREW's successes so far suggest that these plans are not unrealistic. We have already seen, too, that black rhinos in Kenyan reserves breed well. It will be surprising, in short, if captive rhinos in general are not breeding reliably by the late 1990s.

At one time it seemed as if the discussion between the CBSG, and the Indonesia-based biologists, was becoming rancorous. The discussion was indeed caricatured by some outside observers, who represented CBSG as gung-ho interventionists, and the Indonesians as stick-in-mud romantics. Yet at the Rhinoceros Symposium held in San Diego in May 1991, both groups made it clear that they favoured the establishment of a second Javan rhinoceros population – initially in Indonesia. The debate was only about timing and numbers. The committee that was to decide the strategy for Javan rhinos was due to meet early in 1991, but was delayed by the Gulf War. It should convene later in 1991. It will be surprising, though, if captive breeding does not play some part in saving the Javan rhino – though centred, for the forseeable future, in Indonesia.

The fire brigade approach to captive breeding

In the Javan rhino programme (if and when it gets going) only a proportion of the wild animals will be taken for captive breeding, and the captive and the wild herd will complement each other. Often, in the past, captive-breeding programmes have begun much less tidily and (though at the time there may have been no alternative), much less satisfactorily.

Sometimes, captive-breeding programmes have begun by default. The wild (or more or less wild) species has disappeared, and suddenly it was realised that the only individuals left were in some park or zoo. The classic example is Père David's deer – a large, lugubrious, splay-hoofed relative of the red deer. It was once widespread in China. It is versatile, and in the words of one scientist who has worked on it, it is 'as tough as old boots'. But China has long been a hungry and crowded country, and by the end of the last century Père David's deer had been hunted to extinction almost everywhere. Then came the Boxer uprising; and the last few were killed in the Emperor's garden in Peking. (That at least is how it seems; for it is at least conceivable that some survived elsewhere.) But Britain's Duke of Bedford had already taken some into captivity at Woburn. They have bred in Britain and Europe throughout this century – albeit not in accord with the kind of genetic theory outlined in this book, because such theory did not exist until Père David's captive breeding was well advanced. Instead, natural selection was more or less allowed to take its course. Those that survived harsh winters gave birth to the next generation and somehow or other they came through. Some animals, after all, are relatively free of deleterious alleles, and survive genetic bottlenecks; and the Père David's evidently belongs to this fortunate minority.

Now Père David's can be found in captivity worldwide. In New Zealand it is a commercial animal, favoured for venison because of its large size. Britain's farmers would like to farm it too – and perhaps to cross it with the red deer, though the hybrids are not as fertile as is desirable – partly for their large size, and partly because they give birth in April rather than in June as the red deer do, and so their calves are much bigger by the autumn and can be sold for venison before their first winter, like lamb. This book is not about farming, however. The point here is merely that animals *can* be snatched from the grave by captive breeding, and sometimes recover to become positively common. In 1986 London Zoo and Whipsnade Wildlife Park cooperated with Chester, Glasgow,

Longleat, Knowsley Park and Marwell to return Père David's deer to the specially created Da Feng Milu reserve in China. With luck, Père David's deer will be a wild animal (or semi-wild) again. But its survival so far has depended on the collector's instinct – the whim – of a 19th-century English aristocrat.

Several 20th-century captive-breeding plans have begun not so much by default, but as an inescapably necessary last-ditch attempt to save a species from oblivion. Four outstanding cases are the Arabian oryx, the black-footed ferret, the California condor, and the red wolf. It certainly is not ideal for captive breeding to begin so precipitately, with the wild population reduced to the point of disaster. To intervene before that point is reached – as is proposed for the Javan rhino – is obviously preferable. But the last-ditch rescue of the Arabian oryx has become one of the folktales of 20th century conservation, and is worth relating in detail.

How the Arabian oryx was snatched from the grave

Arabian oryx are beautiful creatures, the size of a small pony, with long sweeping horns. The ancient Arabs used to bind their horns together; which, according to Aristotle, gave rise to the myth of the unicorn. This is recorded, too, in *Deuteronomy* 33:17: 'His glory is like the firstling of his bullocks and his horns are like the horns of unicorns.' Arabian oryx are marvellously adapted to the desert; so much so that they once occupied the entire Arabian Peninsula, north into Mesopotamia. But the adaptations that suit them so well to the desert – and the openness of their territory – make them very easy to hunt. They are white, and they deliberately make themselves conspicuous, standing on hillocks so they can be seen by their fellows; and thus they avoid being lost. They are not rapid movers, for that would not be appropriate in the desert heat. They move to new grazing lands by night; and they travel along routes they have travelled before, for that too helps to avoid being lost. They are easy to see, then, and easy to ambush. The Arabs have hunted them from time immemorial.

But so long as the Arabs hunted only from horseback and camelback, and with bows and spears, the oryx had little to fear; and they were plentiful, until well into the last century. Then the British, the Germans, and French brought in rifles, by the million. By the end of the 19th century, very few oryx remained in the north of their range. World War 1 and the Turkish occupation wiped out the oryx from most of Arabia. By 1935 only two populations remained: one in the northern Nafud (on the borders of present-day Jordan and Saudi Arabia); and one in the south, along the border of Saudi Arabia and Oman, in the desert of the Rub-al-Khali.

The final phase of extinction began in the 1950s. The oil industry arrived, bringing vehicles with four-wheel drive, which could follow the oryx almost anywhere. In 1960, Lee Talbot of the Fauna Preservation Society (as it then was) reported that the oryx had gone from the Nafud. All that were left were along the southern edge of the Rub-al-Khali: just a few hundred individuals, which hunters could eliminate in days. The only solution, Talbot felt, was to offer sanctuary by captive breeding. So, in 1961, the FPS launched 'Operation Oryx'; a programme of rescue and captive breeding, with the eventual aim of reproduction.

The details of this rescue are described by Major Ian Grimwood in *The Conservation and Biology of Desert Antelopes*.[6] The time and place to catch the oryx, the FPS was told, was in April and May, on the gravel plains of the north-east corner of the East Aden Protectorate, where they came to escape the extreme heat of the sand seas. But hunters knew this too, and in mid-1961 a party arrived from Qatar. In the space of weeks, they shot 48. The FPS decided to launch its own rescue in April 1962, with the help of the Hadharami Bedouin Legion and the Royal Air Force. But in February 1962 news came of another hunting party, which had killed another 13 animals. Some doubted whether it was worth proceeding at all. Finally, however, the FPS party set out on 23 April.

The FPS searched 8,000 square miles of the Rub-al-Khali, and found evidence of only 11 oryx. Five moved east, over the Oman border, where they could not be found. Two went back into the sand sea. So in the end the FPS caught only four – three males and a female: but one of the males had already been wounded, and died. They had just three, then, with only one female. The RAF flew the animals to Nairobi, and they were held nearby at a specially prepared quarantine station at Isiolo. Three animals (with one female) simply was not enough to begin a worthwhile breeding programme. So the FPS sought animals in captivity. Several zoos said they had Arabian oryx; but some turned out to be addax, and some were scimitar-horned oryx, and some were hybrids. London Zoo did have one, however – a female called Caroline – which it offered for breeding. We now know that at that time there were quite a few privately owned herds in the Middle East, notably in Qatar, so the position was not quite so dire as it seemed; but that was not known at the time. In the event, the FPS located oryx only in Yemen – two animals of unknown sex in the souqs of Taiz; in Kuwait – two females, owned by Sheikh Jaber Abdullah al-Sabah; and in Saudi, where King Saud bin Abdul Aziz kept at least eight in his palace at Riyadh. So it seemed at the time that there were 16 possible founders of the next generation – or rather of all subsequent generations – of Arabian oryx.

Mixed fortunes followed. A definite plus was an offer by the Shikar Safari Club and the Arizona Zoological Society to accommodate the proposed breeding herd at the newly established Phoenix Zoo: a site appropriately named for such a venture, and highly suitable terrain for an oryx. An extra plus was the offer from Sheikh Jaber Abdullah al-Sabah of both his animals, though one died before it could be flown out. King Saud, who was sick at that time, was harder to contact. But he eventually offered two pairs. So it was that four oryx – Tomatum, Pat, and Edith, caught from the wild; and Caroline from London – arrived at Phoenix Airport (via New York) on 25 June 1963. Salwa, the female from Kuwait, followed in September. Edith gave birth on 23 October, having conceived in Isiolo; so now there were six. Caroline gave birth the following spring, having conceived at Phoenix. That made seven. The Saudi pairs arrived in July 1964: Riyadh and Aziz Aziz, the males; and Cuneo and Lucy, females. So now there were 11. In 1964, too, Los Angeles Zoo obtained yet another pair from King Saud, via a Dutch dealer. By late summer in 1964 there were 13 Arabian oryx in two potential herds in North America.

We will pick up on the Arabian oryx story as this chapter progresses. It is a classic, after all; the nearest so far (together, perhaps, with the golden lion tamarin) to the complete sequence of wilderness–zoo–wilderness. Suffice to say here that the last definite report of an Arabian oryx in the wild was in 1972, and though there have been some alleged sightings since then, it is generally assumed that the early 1970s saw the last of them. The point here is simply that modern conservation breeding plans have often not begun tidily. Sometimes the conservationists – in this case the FPS – have swept in like the fire brigade, to a building already on the point of collapse. That is not ideal; not least because the last few animals that found the captive herd seem bound to contain only a small proportion of the total genetic variation of their species. An important task for conservation breeding, therefore, is to establish universal credibility, so that plans can begin before the disaster stage is reached, ideally with animals selected from the wild to span the genetic variation. Meanwhile, we should see the kind of things that happen, and the kind of problems that arise, once the animals are in captivity, and breeding is underway or is at least agreed to be desirable.

BREEDING IN PRACTICE

The way that conservation breeding works is obvious from all that was said in chapter 4. Species coordinators (supported by the studbook data) contrive to ensure that founder populations multiply as rapidly as possible, to minimise genetic loss by drift, and that all the genes of all the founders are well represented in subsequent generations. To this end, they take individuals out of the breeding programme once they have produced their share of offspring – even if (as is quite likely to be the case) those individuals are the best breeders. They seek to equalise family sizes, so that the offspring and hence the genes of individuals that habitually produce large families do not swamp those that produce small families. The sexes should be equalised – same number of males as females; though with the many polygynous species this cuts the animals' social expectations. Matings in general are arranged between animals that are not related to each other (though they should be of the same subspecies). In very small populations, indeed, every male of each generation ideally would mate with every unrelated female – although of course this is rarely achievable in practice.

Genetic variation always is the name of the game. Conservation breeders seek to provide populations that at some future time could return to the wild (if needed to do so), and conservationists are humble souls; they do not presume to judge *which* genes will equip animals for some future wilderness. Animals that breed well in captivity are not necessarily the ones that will do well in the wild. For example, it might well be better in the wild to produce small litters rather than large – and enhance the chances of survival of each individual. If it were not so, then wild animals would habitually produce enormous litters – which is the case in some species but is by no means universal practice.

Yet, again as we saw in chapter 4, a population must contain many hundreds of individuals if it is to be viable in the long term, and retain a high proportion of present-day genetic variation – admitting, though, that present-day genetic variation may be far less than it was 50 or 100 years ago. No single zoo can maintain viable populations of all except the smallest species – and in any case, it is safer if populations are scattered among several institutions. So all these arranged pairings, all these equalisations of litters, all this contraception (or culling!) of over-fecund individuals, has to be a coordinated activity between many zoos. For the projects to work as they should, the species coordinators have to know what they are doing – so they have to be good biologists; the collaborating zoos have to maintain high standards of management, so that they can actually control what their animals are doing (though without oppressing them so that they refuse to breed at all); but the collaborators also have to agree with the spirit of the overall plan and carry out the recommendations of the coordinator.

In the long term, it is in everybody's interests to cooperate, for without cooperation the animals will die out and zoos will disappear too. But in the short term we can easily see that cooperation in a coordinated breeding plan requires enormous self-discipline and some self-sacrifice, as well as breeding and management skills. No curator likes to get rid of (or vasectomise) a beautiful and fecund stud male, or push aside some paragon of a female to make way for some scruffy creature that might happen to contain some recondite allele. No one likes killing cubs; but if a lioness produces five and only two are needed and there is no room for the other three, then something has to go, and if all cubs are kept the whole scheme must rapidly grind to a halt. I know a zoo (it serves no purpose to say which one) that has some of the finest-looking African lions I have seen, that are among the darlings of the visitors. But who needs them? The director and curators of that zoo are conscientious conservators and know full well that they ought to keep Asian lions instead. Only a few hundred Asian lions remain in the world, all in the Forest of Gir in western India, and the European and American breeding populations are having to be started again practically from scratch because it has now been found that most of them contain genes from African lions. But Asians are smaller and generally less impressive than the most magnificent of the Africans. Where will the 'surplus' Africans go? What vet would want to 'put down' such creatures as this?

To be sure, coordinated breeding schemes *can* work. More and more directors and curators are acknowledging that neither they nor a steadily increasing catalogue of species have a future unless they are made to work. The Arabian oryx were extremely well looked after in their early days at Phoenix; they spread from there to San Diego Zoo, and on to Brownsville, Texas; and in 1979 the population spread back to Europe, from Phoenix to Berlin, Antwerp, and Zurich, with the understanding that those animals' offspring would come to London – who had, after all, supplied Caroline, and had proved in the earliest days that they actually recognised an Arabian oryx when they saw one. Arabian oryx have now been reintroduced to the 'wild' (as I will discuss); and there are now about 1,800 Arabian oryx in the world. The species is not entirely out of the woods but it would have to be very unlucky, now, not to pull through. The world's zoos and reserves now contain about 1,000 Przewalski's horses; yet there were only 59 in captivity in 1959, and the horse became extinct in the wild in the late 1960s. In January 1990 there were 305 red pandas in zoos. Many other species and subspecies, from Siberian tiger and golden lion tamarin to Mauritius pink pigeon and Rothschild's mynah from Bali, have been brought from the point of extinction to one of reasonable safety by captive breeding.

The successes are such, and the improvements are so rapid, that only a committed churl could now deride the whole endeavour. Captive breeding is worthwhile, and it can be made to work. But if it is truly to succeed then we must acknowledge the shortcomings. These, then, I will focus upon; not in a spirit of negativity; but simply because (as Napoleon observed) it is best to know what the problems really are.

Problems – logistical

One basic problem is social/diplomatic/organisational. Zoos are run in very different ways (municipal, national, some as charities and some as private enterprises), and to some extent we should say *vive la différence!* for every system has its advantages and disadvantages, and what counts in the end is that good individuals should occupy the key positions. This, however, is not always the case. Some zoo directors are simply public servants and by the time they have learnt the rudiments of conservation it is time for them to leave; while others learnt all they are going to learn 20 years ago and don't intend to change now, and be told what to do by a species coordinator. As in all aspects of life, expert management is needed, and continuity without fossilisation; but a significant task at present for zoo conservationists (which with luck will fade away as the years pass) is to educate those who do not yet realise what is going on. To be fair, too, it is hard enough to run a zoo even without the added burdens of coordinated breeding; and the other problems (finance, building regulations, fire insurance, outbreaks of disease, breakouts of animals) often tend in practice to seem more pressing.

In addition, curators find many legitimate reasons for resisting the plans of species coordinators. There are innate conflicts. A stud male or an old female who have already spread their genes widely enough may none the less make excellent and necessary contributions to the social group. A male may be vasectomised – but if he is left with the herd, he may still prevent other males from mating. An old female might be left as an 'aunt', but then she occupies ever-precious space. In zoos, it is at least relatively easy (usually!) to exercise control – for example to catch an animal for vasectomy by darting it. But when animals are returned to the wild the conflicts between social and reproductive needs may become more significant (because the social input of the old experienced animals becomes more necessary); yet at the same time the management of reproduction becomes more difficult. Thus it is often suggested that management in the wild should probably be on a sub-population by sub-population basis, rather than at the individual level. If the wild population had a good variety of founders, and is reasonably numerous, then that ought to work.

Curators care, too, about the welfare of individual animals, as indeed they must. Thus schemes for breeding gorillas have at times been held up because curators were reluctant to allow individual males to travel to new groups – very reasonably arguing that this would upset the male himself, the group he has left behind, and the group he is going to. Time was when zoologists warned against the mortal conceptual sin of anthro-pomorphism (see chapter 7); of seeking to understand animals by looking at their world through human eyes. But it is certainly reasonable to empathise with a gorilla and his family, when he is summarily extracted and dumped among strangers. Artificial insemination (chapter 6) will help in many such cases; though AI of exotic animals requires handling under anaesthetic and, in practice, gorillas are not yet among the species for which AI can be successfully carried out.

There are veterinary restrictions, too, on the movement of animals, sometimes between countries, and sometimes between states of countries. Britain, for example, imposes six months quarantine for rabies for many mammals; many additional restrictions for hoofed animals; 35 days for birds; and so on. It seems foolish to object to such obvious precautions; yet one does get the impression, in other contexts, that money talks. The restrictions on hoofed animals are largely waived for racehorses. Lambs hurtle across Europe in enormous wagons and often hideous conditions for no reason except that some people find it profitable. Meanwhile, the restrictions that should be obviously applied to intact animals may also be applied unwaveringly to frozen gametes (chapter 6); so even when AI is technically as proficient as it surely will become, and could be used to save most animals an unwelcome trip, it still might not be applicable in practice. Inevitably, then, veterinary restrictions complicate international and some inter-regional breeding projects; and some curators who would be prepared to allow animals to make quick visits to zoos nearby are reluctant if this involves quarantine.

Even when plenty of people are cooperating, however, and there are plenty of animals on board, there may still be problems that have to do with the animals themselves – or rather with the past history of those animals. Because new breeding groups are often – in practice usually – founded untidily, from those who were available at the time the decision to begin was made; because curators in the past did not appreciate the modern techniques for ensuring genetic diversity (or the need for it); and because the differences between subspecies were sometimes not appreciated, present-day breeding populations sometimes have a lot of short-comings. None has more than Przewalski's horse.

The troubles of Przewalski's horse

Prezewalski's horse – stocky, pony-sized, big-headed, with a mane like a scrubbing brush and apparently very similar to the ancestor of the domestic horse – provides a prime example. At the end of 1985 the captive population numbered 660, in North America, Continental Europe, Britain, and Australia, and people were already thinking about reintroduction to Mongolia. But the present captive herd can trace its origins back to the turn of this century, and – with no modern theory for guidance, and a certain amount of bad luck – there has been plenty of time for things to go wrong. In consequence, the present population needs weeding.

Thus, for example, curators have tended to keep Przewalski's horses in harems, for that is how they live in the wild, and what is natural is very reasonably perceived to be good. But as Lydia Kolter and Waltraut Zimmermann of Cologne Zoo point out in the *EEP Coordinators' Manual* of May 1989 (Dr Zimmermann is the EEP Przewalski's species co-ordinator),[7] such social structuring breaks the cardinal genetic law which says that for the sake of genetic variation, males and females should be equalised. Yet breeding in harems has been going on for decades, for the present animals have been in captivity for between five and eleven generations. Worse is the conflict between present theory and what seems to be common sense. Thus, theory says that all founder stallions should be encouraged to breed; but former curators (following the well-established precedent of domestic horse-breeders) deliberately chose to breed only from the best breeders. That does seem sensible; but it is ruinous, if genetic variation is the goal, and genetic variation must be the goal if we truly want to conserve *wild* animals, and return them to the wild.

The consequence of these historic policies is twofold. First, much valuable genetic variation has been lost; indeed the present herd had only 13 founders, and the genes from two of them have all but disappeared. Secondly, many of the remaining animals are severely inbred, and hence are highly homozygous. This is shown by the studbook histories; it can be inferred just by looking at the animals – for in some herds the animals all look much the same; it can be perceived nowadays through studies of DNA; and in some animals it manifests in infertility (nonviable sperm). Potentially serious, too, is the nervous disorder that manifests in lack of coordination – ataxia – which occurs in some herds; and which may be genetically transmitted.

Then there has been bad luck, dashed with a measure of carelessness. One of the 13 founders of the present herd, at the beginning of this century, was a domestic mare: *Equus caballus*, as opposed to *E. przewalskii*. Later, in the fourth generation, another *caballus* mare was crossed with a Przewalski stallion, apparently to see how closely the two species are related. One of their hybrid sons remained in the breeding herd. It seems quite likely, too, that wild Przewalski's that roamed the Mongolian plains in the 19th century also had some domestic horse blood. It would be surprising if they did not, for the Mongolians are archetypal horse people, and there surely must have been some inter-breeding. Some early photographs of wild Przewalski's suggest a hint of *E. caballus*.

Be that as it may, some modern Przewalski's do look remarkably like domestic horses; and, in particular, an increasing number – reaching 3 per cent by the mid-1980s – are born with the colour known as 'fox', which horsy people would call 'chestnut'. Foxiness is brought about by lack of black pigmentation, controlled by a recessive allele. Analysis of pedigrees show that at least three of the founders must have carried the allele, but scientists and hunters who described Przewalski's in the wild never commented on such a colouring. Fox-colour, then, can be taken as a domestic trait. Many animals in the British and European herds carry the allele, because two of the founder stallions of those herds were carriers. One of the two sired more than 100 foals, of whom at least 50 (by the rules of Mendelian inheritance) could also be carriers.

If Przewalski's horse still existed in the wild, then the species coordinator might well decide to do as the breeders of Asian lions did, when African genes were discovered among them: start again. But Przewalski's is now extinct in the wild, and that is not an option. Instead, say Drs Kolter and Zimmermann, the plan must be to tidy up the present herd: reduce the inbreeding; select to reduce the influence of the domestic horse; increase the pool of stallions, so that there are more to select from; and to select to reduce health problems such as ataxia.

The precise, exhaustive details we can leave to the coordinators and curators, but a few salient points are worth noting. First, it is most unusual for breeders of endangered species to select deliberately to emphasise or eliminate some phenotypic character: that is, some character that is actually visible. Normally such selection is the prerogative of farmers and dog fanciers. But this is a special case. It seems perverse not to seek to eliminate a character – foxiness – that is known to have come from another species. It is not entirely unprecedented among conservation breeders to seek to eliminate genes that are known to be deleterious (in this case the hypothetical gene for ataxia). But Przewalski's is a special case. We should be grateful that the animal still exists at all, though glad that there are none others that present quite such a range of problems.

Second, the call for more stallions raises the issue of bachelor herds (as mentioned above). Among many wild animals, the sexes separate for much or most of their life; and in many, the young males naturally congregate (as indeed is the case among humans). Zoos that keep males and females with youngsters at heel are obviously breeding animals, and invite public approval, while baby animals are always attractive. But in the field of conservation breeding, they also serve who also do other things; and zoos that keep bachelor herds for other zoos to draw upon can do a very useful job. Cologne is among the European zoos with such a herd (and very fine they look, too). Washington D.C. incidentally, has pleasant bachelor groups of ruffed lemurs. It is particularly important for zoos that keep such groups to explain what is going on, however, or the visitor thinks that the animals are simply being kept for the sake of it; and as bachelors can make an enormous fuss as they try their strength (*vive* the Washington lemurs) the impression is falsely gained that the whole exercise is gratuitously unkind.

Third, there are some Przewalski's horses in the Soviet Union, which apparently contain genes that are not present in the coordinated herds. The sooner these are brought into the fold, the better. The same applies, for example, to the red and giant pandas of China. Genetic exchange is vital for the survival of all. Finally, the numbers of Przewalski's are now such that a return to the 'wild' – or the modern equivalent thereof – can now be contemplated. After years of negotiation and search (it really is remarkably difficult to find a suitable site for the reintroduction) the Mongolian government has allocated 100 hectares of the Gobi-Alti national park, which will be progressively expanded as the animals grow used to their new surroundings. The reintroduction is scheduled to begin later in 1991; September is the scheduled date at the time of writing. I will discuss the general problems of reintroduction later.

Problems – space

Lastly, overshadowing and complicating all captive breeding endeavours is the perennial problem of space. Species coordinators know, as a generalisation, that the total zoo population of their particular protégé can never exceed a particular number, because all the zoos that they deal with between them can allocate only that number of spaces. One of their first tasks as species coordinators is to ascertain what that 'particular number' is. Organising breeding to maximise genetic diversity is only half of the problem. The other half is to ensure that that diversity is maintained within the population that can, in reality, be kept. Sometimes the figures do not add up. Sometimes there are too few spaces for the task in hand.

This problem is of course exacerbated if the species is divided into subspecies; and a case in point is the red panda. This is one of the showcase animals, which demonstrates the efficacy of captive breeding. The first international studbook for the red panda, published in 1978, registered only 128 animals (though the true figure in captivity was probably 143); while, as noted above, there were 305 on 1 January 1990. The red panda's studbook keeper, Angela Glatson of Rotterdam Zoo, Netherlands, notes in the *International Zoo News* of September 1990[8] that more than 100 have been brought in from the wild in that time, so the expansion is not quite as impressive as it may seem; though she also notes that whereas only a third of the captive animals of 1978 had been born in captivity, almost 100 per cent of the nominate subspecies, *Ailurus fulgens fulgens*, were born in captivity.

But herein lies the real snag. As noted in chapter 4, there are two subspecies of red panda: *A.f.fulgens*, and the Chinese or Styan's red panda *A.f. styani*. They are genetically and phenotypically distinct, and they should be raised separately. The present 300 or so captive red pandas comprise about 120 Styans and 180 nominates. The present *fulgens* population effectively has 26 founders; the generation length is five years; and the kind of calculations outlined in chapter 4 suggest, therefore, that the population required to maintain 90 per cent of present genetic variation for 200 years would be 500, which is well over twice the present figure. There should also be 500 Styan's. So the total should reach 1,000, as quickly as possible, and preferably by 2000 AD. But the world's zoos may not muster so many places between them. Australia (which has a coordinated breeding plan) apparently has already reached capacity. What a pity if the two subspecies were effectively thrown into competition with each other, and thus produced two populations that were both inadequate!

Increasingly it will be necessary to *make* more spaces for the most endangered animals by getting rid of (which will often mean culling) animals that are safe in the wild and do not require captive breeding, or are being bred elsewhere, or are hybrids. On the very day I am writing this, members of the Primate Specialist Group of the Federation of British Zoos are meeting to discuss the ideal space allocation for all primates in British zoos, from lemurs to gorillas. Some ideally should not be kept at all, they feel, because the Americans are already breeding them. Others should go because they are common in the wild; while some populations should be rapidly expanded. Even before the reallocation can begin, however (assuming the curators agree), there has to be a radical sorting out; for among monkeys in particular (as we have noted already among owl monkeys) the scope for confusion between subspecies or even between subspecies, is acute. For example, the four recognised species of spider monkey (genus *Ateles*), have 16 recognised subspecies between them. Roy Powell of Paignton Zoo in the west of England is the regional studbook keeper for spider monkeys. Just to take one example, Dr Powell noted[9] that the black-faced black spider monkey (*A. paniscus chamek*) from Peru, Bolivia, and Brazil is very similar to the robust black spider monkey (*A. fusciceps robustus*) from Colombia and Panama, which is critically endangered; and, of course, the resulting hybrids contribute nothing to conservation even if they are fertile. More hybrids. They are nice animals, but they are definitely getting in the way.

Culling cannot be avoided; and probably it is foolish to be squeamish. We kill thousands of animals every day for food, thousands more because we regard them as pests, and millions more out of insouciance, as we flood and hack their habitats. A cow or a rat think no less of themselves, presumably, than a hybrid spider monkey. Even so, everyone would agree that killing should be kept to a minimum; at least, it seems to be against the spirit of the endeavour. Here, then, perhaps, is a case for the animal sanctuary; places where animals can live out their lives when they are not wanted for breeding. Such places could be anywhere that is secure, and as easily run by responsible amateurs as by professionals; suburban gardens, hotels, farms.

I will return to the role of the amateur and the freelance in chapter 8. First it is relevant to discuss two categories of animal that are not generally thought of as prime targets for captive breeding, and yet have many advantages – and so could be of special interest to amateurs. These are fish and invertebrates.

FISH AND INVERTEBRATES

I hope fish, invertebrates, and their various advocates will forgive me for lumping them together; but for conservational purposes they have much in common. First there are the sheer number of species. Twenty-four thousand species of fish have been named, but among teleosts alone (which are the chief group of bony fish) 100 new species are named each year; and there may in reality be 35–40,000 species of fish. The number of invertebrates is even less certain; but as we noted in chapter 2, 30 million would not be an unreasonable estimate. It is difficult enough to cobble together the space to breed 2,000 land vertebrates effectively. Fish and invertebrates seem way beyond practicality.

There is the matter, too, of specialisation. The ultimate purpose of captive breeding is to return animals to the wild. Most of the land vertebrates that are being bred in captivity are reasonably versatile, and many, such as the Arabian oryx, once had huge ranges. They can be kept, at least temporarily, in reserves that may be far from their native lands. When they are finally returned to the wild, it may be to areas that are separate from the ones in which they last lived, but are none the less part of their original range. Arabian oryx did once live in Saudi Arabia, Jordan, Oman, and Israel, where they have been returned. The California condor could well be put back into the Grand Canyon rather than California; but it lived there once, so the fossil record attests.

The habitats of fish and many invertebrates, on the other hand, are often extremely localised. There is a huge variety of freshwater fish largely because the streams and ponds in which each one lives are cut off from all the others – unless, for example, a stream-living fish is also able to survive in the big river, and so enter another tributary, which is often not the case. Many insects live only on particular plants; or, like the Large Blue butterfly that once inhabited southern England and is now extinct, they relied on particular combinations of plants and other animals. If particular streams are drained (or flooded by dams), or particular plants disappear, then there is nowhere for the captive-bred animal to return to. Sometimes, to be sure, a fish from a pond that no longer exists could be returned successfully to some other pond. But what effect would this have on the fish that are native to that new pond? Huge numbers of fish have been annihilated or severely endangered worldwide precisely because meddling anglers and fish farmers have put novel species in their midst; zander into Britain, eating the roach; trout into the southern hemisphere, wiping out many of the southern equivalents that fill the same niches; rainbow and hybridised farm trout into British rivers, compromising the native brown trout; big carnivorous bass and catfish into warm waters worldwide, eating everything; Nile perch into Africa's Lake Victoria. The only encouraging suggestion I have heard came from Dr Peter Maitland of the Fish Conservation Centre at Stirling, Scotland, made at a symposium at the University of Lancaster in 1990:[10] that rare fish from captivity might be returned to new waterways, such as reservoirs, to create quite new ecosystems. On the whole though, the criticism that is often levelled at captive breeding in general – that habitats disappear, and leave captive animals without a home to go to – seems to apply more often to fish and invertebrates than to land vertebrates.

I have mentioned that the Nile perch, *Lates*, has harmed the native fish in Lake Victoria. That is an understatement. Since its introduction in the late 1950s it seems to have driven to extinction about 200 of the 300 different species of haplochromine fish – small members of the family Cichlidae – that formerly lived there. Nile perch is a good food fish and economically it brings a net advantage – although it has economic disadvantages as well. But ecologically it has wrought perhaps the greatest disaster of the 20th century. Only three aquaria in the world had specimens of any of the 200 haplochromines that were rendered extinct, one of them being Horniman Museum in South London where Gordon Reid is keeper of natural history. Dr Reid maintains nine species of haplochromine. The museum is not big: it was founded as a private collection for public benefit by a philanthropic tea baron in the 19th century. But several of the species at present live only at Horniman, in tiny tanks; and *nowhere else in the world*. Similarly, Philip Pister of the California Department of Fish and Game, at Bishop, likes to recall the time he held an entire species of desert pupfish in couple of buckets.

What should be done with such fish, that have nowhere to go? It really does seem a shame simply to allow them to die. Philip Pister did find somewhere else for his pupfish to live and Gordon Reid has founded the Fish Rescue and Breeding Centre at Horniman, under private sponsorship, and under the aegis of the International Association for Research on and Conservation of Endangered Cichlids (IARCEC). However, Reid is fully aware that no single institution can last forever, so no population should be held indefinitely in only a few tanks; that the fish populations must be expanded rapidly, in accord with genetic principles; but that Horniman itself is of very limited size, and cannot accommodate huge populations of several species.

The obvious solution, then, is to elicit the help of amateur aquarists, who are willing to spend hundreds of hours of their time and collectively spend an estimated £7 billion on their hobby worldwide. Which brings us to the huge innate advantage of at least some fish and many invertebrates over most land vertebrates. Many are extremely small, and easily contained in tanks. Large populations can be kept in confined spaces cheaply; and can as soon be kept by enthusiastic amateurs as by professionals. Similarly, Paul Pearce Kelly, head keeper of invertebrates at London Zoo, has at least contemplated the possibility of inviting amateurs to take on gangs of red-kneed tarantulas, which he produces by the thousand in captivity (and all have to be individually fed) though they are endangered in the wild.

In short, captive breeding cannot make a huge quantitative impact on the plight of fish and invertebrates. But it can save at least those that happen to cross our path; and it can do so relatively cheaply, and indeed could defray much of the cost onto hobbyists. People who now keep guppies and mutant koi could, if they chose, become significant conservators of entire species.

As we have constantly emphasised, however, the proper end point of captive breeding is reintroduction. This we should now discuss.

BACK TO THE WILD

Probably only a minority of populations bred in captivity will be returned to the wild in the foreseeable future. For some, return may not be possible until many more decades have passed. For others – many, we may hope – mass return may never be necessary, because the wild populations might yet be saved. Wild populations are harder to manage than captive populations, however, and even the 'safe' ones may need genetic support (for future populations are bound to be smaller than those of pre-human times). Neither can we foresee all that fate holds in store for the wilderness of the future, and captive populations should always be capable of return to the wild, if called upon.

In reality, too, reintroductions do not have to be absolute, in order to be useful. That is, in the three most famous cases – Père David's deer, Przewalski's horse, and Arabian oryx – the animals were almost certainly extinct in the wild, and survived only because they had been bred in zoos and parks. The ones that were put back were the only non-captive individuals in existence. But reintroduction can be worthwhile – and in some cases vital for species survival – even when the species does still exist in the wild. Thus, as discussed in more detail later, golden lion tamarins are being reintroduced into Brazil, and Rothschild's mynah (alias Bali starling) into Indonesia, not to restore an extinct animal but to replenish wild populations that have been dangerously depleted. We may note in this context that captive populations are often more diverse genetically than wild populations of equivalent size, because the captive populations may well have been founded by wild individuals from many different locations, and because nowadays they are bred specifically to maintain genetic diversity. Thus animals introduced to the wild from captivity can be a particularly potent source of genetic variation. To put the matter another way: there may only be a few *species* in captivity that are extinct in the wild; but there is undoubtedly a huge array of *genes* among captive animals that are already extinct in the wild. If those genes are to be returned to the wild populations, then the individuals who carry them must be capable of survival in the wild (unless, of course, wild females are impregnated by AI using the sperm of captive animals, as described in chapter 6).

There are many schemes, too, not to reintroduce animals that are totally extinct in the wild, but simply to restore them to parts of their former range where they are now extinct. Thus in Britain, at the time of writing, the Royal Society for the Protection of Birds and the Nature Conservancy Council are reintroducing the red kite to England and Scotland. This bird was once common and widespread in Britain. It feeds on carrion and small mammals, and in less hygenic days there was plenty about. London offered especially rich pickings, and the kites (it is said) were as thick in the air as is their equivalent species in present-day Karachi. They declined as street cleaning became fashionable, and gamekeepers began to wage war on them in their inimitably cavalier fashion, until the kites remained in only a few places in upland Wales. They survive in Spain and Scandinavia, however, and are being reintroduced from there. To be sure, this in large part is an exercise simply in 'translocation'; wild animals are simply being moved from one part of their range to a part they no longer occupy. But some birds are being brought in specifically for breeding (including some from a sanctuary in Spain that have been injured, and cannot be released). Thus, captive-bred red kites will swell the ranks of the wild; and thus we see again that reintroduction is not absolute, but can involve many shades of endeavour.

Reintroduction is never easy, however. Experience gained through various endeavours throughout this century has now enabled IUCN to lay down criteria and guidelines. Paramount among them are that the new site should be thoroughly researched. If a particular site last harboured the particular creature 10 years or 100 years, or 1,000 years previously, then, very probably, it will have changed in the intervening years. Does it still provide all the animal needs – nest sites, prey, water, whatever? We know, for example, that the Middle Eastern deserts have been changed radically in recent decades by various 'improvement' schemes, including construction of artesian wells. Is it all still fit for oryx? Primates – orangs, tamarins, woolly monkeys – may be returned to forest; but present forest is likely to be secondary forest (the kind that grows up after primary forest has been felled). Is it suitable?

Even if the territory has not significantly changed since the animal last occupied it, we must ask – why, then, did the animal become extinct in the first place? Do the same dangers that drove it to extinction then, still exist? Two of Britain's reintroduced red kites have already been poisoned by gamekeepers, and although we should still be hopeful (for public opinion is on the side of the birds these days, and those cases led to prosecutions) we have to admit that the initial danger has not entirely gone away. It is indeed vital to ensure (another IUCN stipulation) that the local people are ready to receive the reintroduced animals; and far preferable to find ways in which the local economy (as well as general quality of life) can benefit from their return. As we noted in chapter 1: direct benefit to human beings should not be seen as the *point* of conservation; but conservation projects that can bring some benefit are far more likely to succeed. We will see this principle in action in the examples below.

Finally, it is at least highly desirable that animals should be returned only to areas that they inhabited in the past. Otherwise, they are likely simply to be interlopers in some other animal's territory – and as we saw in chapter 2, animals (or plants) introduced out of the blue from some alien territory are one of the principal causes of extinction (among the native animals). Sometimes this point raises some intriguing dilemmas. For example, Mark Stanley Price, director of the African Wildlife Foundation at Nairobi, has been involved in relocating black rhinos in the Lake Nakuru National Park Kenya.[11] One way to finance this return is to present the animals as a tourist attraction. However, black rhinos are woodland creatures (solitary browsers) and are not easily seen; so the paying visitors might feel cheated. One way around the dilemma would be to introduce white rhinos as well; for these graze in open country, and tend to move in herds. Most tourists (who are generally in a tremendous hurry in any case) would not mind too much about the species difference. However, the palaeontological records suggest that white rhinos have not lived in the Lake Nakuru area for several million years. They do live in Kenya, to be sure. But they are not native to this particular location. How purist can we afford to be? Besides, Lake Nakuru has all the large grazing animals it needs which are truly native to the area. If white rhinos come, something else will be compromised. So should we say: Lake Nakuru is near enough to the white rhinos' natural habitat, so let's go for it? But if we did say that, then where would we draw the line? In this case, it really seems that the vote should go the white rhinos – introduced to take the pressure off the black. But we can see that the dilemma is real.

As we observed in chapter 2, we would not expect reintroductions to go into full flood until several centuries hence, when the human population has again begun to fall; and, as noted earlier in this chapter, we cannot expect that all captive populations can be returned in the foreseeble future, or will need to be returned *en masse*, even in the distant future. But already, despite the caveats and constraints, about 100 reintroductions are underway worldwide, or are already well established. I will look at a few, which illustrate the principles, the difficulties, the variety of undertakings (for every instance is different) and – sometimes – the possibility of triumph over adversity.

The rescue of the Arabian oryx from oblivion is a wonderful tale of 20th-century conservation. So is its return to the wild.

The return of the Arabian oryx

G. Laycock in *The Alien Animals*[12] quotes a contemporary account of an attempt in the 1870s to introduce a mixed bag of some 4000 European songbirds into a suburb of Cincinnati: 'A cloud of beautiful plumage burst through the open window, and a moment later Burnet Woods was resonate with a melody of thanksgiving never heard before and probably never heard since'. I stole that quote from the excellent *Naturalised Birds of the World* by Sir Christopher Lever, who comments: 'Few of those birds were ever seen again.'[13]

But in the words of Mark Stanley Price: 'There's more to reintroduction than opening a box and allowing the animals to go.'[14] He should know, perhaps better than anybody; for before he took over the African Wildlife Foundation he and his wife spent seven years in the desert, organising and managing the reintroduction of the Arabian oryx to

Oman; to date perhaps the most spectacular and complete reintroduction yet achieved.

The reintroduction phase of the Arabian oryx story began at the end of the 1970s by which time, as we have already seen, the animals were breeding well in various zoos in the United States and had been redistributed to zoos in Europe. The first returns to the Middle East were not in practice to Oman but to Jordan, in 1978; from San Diego to the New Shaumari Reserve. But Oman's remains the classic reintroduction story. The Sultan of Oman, seeking in general to study and protect the much-beleagured wildlife of his country, felt that the return of the Arabian oryx would be of especial significance both ecologically and symbolically. Ralph H. Daly was appointed Adviser for the Conservation of the Environment; and initial ecological studies suggested that no matter how unpromising Oman might seem to non-desert creatures such as human beings, it was still fit for oryx. The site finally selected for the initial reintroduction was Yalooni, on a plateau known as the Jiddat-al-Harasis; about 150 m above sea level and on the very edge of the tropics, about 20° north of the Equator. The hottest June days at Yalooni reach 47° C, while the coldest Januaries average only 7 degrees, and each day the temperature fluctuates through 15° to 20° C. The little rain that falls on the Jiddat-al-Harasis (0–200 mm per year, averaging around 50 mm) does not stay long on the surface, and there is no other surface water. Moisture is supplied most reliably by fogs, rolling in from the Arabian Sea. These supply about 18 litres per square metre per year, mainly in autumn and spring; and thus there are two good growing seasons, and abundant vegetation, with grasses and various leguminous trees including acacias and *Prosopis*, a relative of the Mexican mesquite tree. Perfect oryx country.

In 1980 18 animals finally arrived at Yalooni for reintroduction (though not all together). Seventeen came directly from the United States, and one from Jordan (which had previously come from the US). The procedure for reintroduction in essence then followed what has become the standard protocol for reintroduction. In captivity in America, the released animals (and their parents) had been kept in territory comparable to that of the desert; southern California (San Diego) and Phoenix, Arizona. At Yalooni, they were first kept in pens, 20 by 20 m; and the pens were in a larger enclosure, one km square. Thus confined – or semi-confined – the animals could acclimatise, and be studied. Yalooni thus became a reserve, a research centre, and the base-camp for the newly wild animals – or at least, for the first generations. Thus, the return to the wild is not precipitate, but by graded steps. Always there is the opportunity to retreat, regroup, and try again. Each step is obsessively monitored.

The tribespeople of the region are the Harasis. They, with their Sultan, welcomed the opportunity to take the oryx under their wing, and to look after them. Their help has in practice been vital, as protectors, wardens, and observers; and they in turn have benefited from the project, for they have earned more from it in wages through the 1980s than from the oil industry. The oryx have again become part of their culture, and are the major prop of their economy. Thus we have a perfect example of the point made in chapter 1: that conservation projects can contribute to economies; that they must contribute to economies (or at least not detract from them) if they are to be secure; that they must have the approval and connivance of the local people; but that benefit to human beings should not be the prime motivation for those projects. The Sultan and the Harasis did not expect to grow rich from the oryx. They just felt that the oryx was a part of them.

Yet at the very beginning, the Harasis of Yalooni were sceptical. They did not believe that zoo-bred animals could be the same as the oryx they remembered, or that they could survive. Yet as Ralph Daly recalls in *The Conservation and Biology of Desert Antelopes*,

. . . when the oryx stepped out of their crates into the pens at Yalooni on 10 March 1980 . . . For the old men it was a return of . . . part of their past: for the younger ones . . . the materialisation of a mystical element of their cultural history.

From that day onwards, the Oryx project with its centre at Yalooni started to become an integral part of the life and economy of the Harasis.

Incidentally, at the time of their release the animals had been in their individual crates for 45 to 90 hours; but they showed no signs of stress from lack of water.

Acclimatisation of the individual animals to Yalooni – to climate, diet, and the smells and sounds of vehicles and people (as well as of sand and acacias!) – was one initial requirement. Another, as Mark Stanley Price recalls, was to ensure that the animals formed a coherent social group before they were released. Oryx are social, and normally live as part of a coherent herd. They rely upon each other. They can survive alone, but they do better in a group. But if they are to stay together, the members of the herd have to get along with each other – just as is true of humans. Each has to know his or her place in the hierarchy or they will squabble. The herd has to have a good leader: intelligent, not too nervous, but not too trusting of human beings either. Both for social and for breeding purposes, there has to be a proper ratio of sexes, and preferably of ages. Dr Stanley Price and his colleagues felt that the herd to be released should be structured like the typical herds of former times: at least ten individuals, with about equal numbers of males and females, who had been together long enough to establish a stable hierarchy. In the end, two of the animals proved unsuitable for release: one old male who paced all the time, and one young female who was too attracted to people. These two should have been left in America.

Mark Stanley Price felt, too, that after the oryx were released they should recognise Yalooni as a home range, and should develop some fidelity to it. The animals would have a greater chance of survival if they knew they could return to base if they got into trouble. Note, throughout, the blend of detached biology – what are the needs of the animal? – with ostensibly anthropomorphic common sense (see chapter 7). If you or I were being let loose in the desert, we would like somewhere to return to, too. So why not an antelope?

Finally, Stanley Price and his colleagues determined to collect as much information as possible about the behaviour and adaptations of the animals before release, and for as long as possible afterwards. This they did; and much of what is now known about the Arabian oryx (and indeed about adaptation to the desert in general) derives from those studies. The observations were made largely by the Harasis rangers, who tracked and watched the animals every day after their release. They did this by fitting some of the animals with radio collars – and by this means the location of almost every animal on every day was plotted until at least 1986. The Harasis initially practiced the necessary techniques with the help of a radio-collared camel. Radio collars, together with field-glasses, tele-scopic camera lenses, video cameras and anaesthetic darts, are among the technologies that have transformed the study of animals in the wild over the past few years.

Most of the animals spent two to seven months in the pens, and from two to 24 months in the outside enclosure. They had access to water at night, and were fed hay and lucerne (alfalfa). Eventually, they formed the stable hierarchy that was considered so essential; they all moved around together, and the lead male herded the rest, and marked out his territory by defaecating in a characteristic squatting position – a sign that he felt himself to be in charge. When this stage was reached, it was necessary only to wait for the hottest season to pass, to effect the release. Sixteen of the original 18 animals were fit for release, plus calves; so Stanley Price felt there were enough to form two herds. Accordingly, Herd 1 was released on 31 January 1982, with four males and six females, of whom two had been born in the enclosure; and Herd 2 was released on 4 April 1984, with four males (one born in the enclosure) and seven females (three born in the enclosure).

Subsequent observations on the oryx (largely by the Harasis) confirm that their adaptation to the desert, and their ability to exploit what seems such an unpromising environment, are truly wonderful. More to the point, although the animals needed to find their way around the particular territory (for as I will discuss later in this chapter, and again in chapter 7, animals need to *learn* far more than has hitherto been recognised), it was clear that captivity had not significantly blunted their desert instincts. For example, the oryx can do without surface water entirely except when pregnant, provided only that the vegetation is reasonably moist and the temperature is not too high. But they do make as much use as possible of fog, by licking stones (or each other) as it condenses. They also know when to seek shade. Their ability to map their terrain and bring it within their terrain is even more remarkable than their simple ability to endure. Four years after release, Herd 1 had a total range that covered 2,000 square km – and by 1986 this was up to 3,000. Their territory was not in one block, but was a series of favoured grazing areas, each between 35 and 350 square km.

They explored their environment (and thus built up their mental maps) by a simple strategy. They grazed an area until they smelt fresh grazing at some distant place. In other words, they stayed in one place until it rained somewhere else and stimulated plants to grow; or until the wind shifted, and brought them news of pastures elsewhere. They travelled by night, commonly walking 45 km. Once they had been to a place, and been rewarded, they remembered where it was, and could find it again from any direction. You may doubt that this is possible, but one particular female demonstrated the point. She was pregnant, and needed to drink. So one night she set off and walked to the base at Yalooni – the only place on the entire Jiddat-al-Harasis where water was always available. Her tracks showed that she walked unerringly to the trough, even though her last visit had been more then nine months before. She drank 11 litres – for when they do take time off to drink, oryx can consume up to 20 per cent of their body weight. Then she returned straight to the herd. Their sense of direction is one cause for wonder. Another is that they can judge whether it is worthwhile to undertake a particular journey for a particular reward. Herds as a whole commonly revisit sites that they have not been to for months or years; and when they do, they may home unerringly upon some favoured spot, like a heap of camel bones, that they had chewed upon before.

Thus, though four of the original 18 immigrants had died by mid-1987, none that was past weaning had been lost, or died from thirst or starvation. A snakebite killed one of the casualties; and two females, who were keen scavengers, died from botulism. Twenty-nine calves were born in those early years, of whom three were stillborn. Five died before they were 30 days old, which is the same rate of mortality as has been recorded for hoofed animals at Whipsnade. One was killed by a mass attack of ravens (which also feature in the Old Testament as a hazard for lambs). Life is even tougher in the great outdoors than we might care to imagine.

The problem, though, was and is to avoid inbreeding. It is theoretically possible (expensive, but theoretically possible) to manage the breeding of wild animals to some extent (for example by removing particular males that have sired too many offspring). But it is not possible to fine-tune the breeding. In practice, the reintroduced Omani females did not have equal numbers of calves; which alone must increase the tendency to inbreeding. Death in the first 30 days of life is a sensitive indicator of inbreeding depression – and in general, in Oman, the ones which theory says are the most inbred (with the highest inbreeding coefficient) have fared the least well. Intriguingly, too, the ill effects of inbreeding were evident at lower inbreeding coefficients than in zoos – suggesting that in the wild, the selective pressures are higher.

To some extent, however, the animals' social behaviour tended to reduce inbreeding. Thus, the two herds that were released separately met up in mid-1984, soon after the second herd was released. Over the following six weeks, many animals changed herds, either singly or in small groups. At the end of that time, only four animals were not in their original groups, and of those exchanges, only two – both males – proved permanent. But such an exchange would reduce inbreeding while maintaining social coherence.

Encouraging, too, has been the tendency of the herds to break up as the numbers grow (15 per herd seems the maximum for comfort) and as individuals gain in confidence. By early 1986, small temporary sub-groups began appearing in both herds; and by June of that year, when heavy rain attracted all the animals to an area in the south, they moved not as two discrete herds, but in several small groups. Provided each small group contains knowledgeable individuals that can find their way about, it makes sense to keep the groups small. It is better genetically (provided the small groups come together from time to time) and ecologically, because a small group can survive on a smaller area of grazing than a large one can.

Now, 1991, the Omani herd continues to thrive, and there are other herds in Qatar, Abu Dhabi, Jordan, Saudi Arabia, Bahrain, and Dubai, which continue to expand. These animals include those from the US-derived World Herd, plus others (including those from Qatar) that survived in parks (unknown to the Operation Oryx biologists). Mark Stanley Price has moved on to the Kenya Wildlife Service; but he continues to warn (for example at the Zoological Society of London's symposium 'Beyond Captive Breeding' in November 1989) that unless the animals are monitored and managed, then genetic bottlenecks leading to inbreeding can still occur. David Jones of London Zoo argues, in *The Conservation and Biology of Desert Antelopes*, that a major and urgent task now is to integrate the genes from the animals that were not in the US World Herd, including those from Qatar. Scientists from the Zoological Society of London are now complementing the monitoring efforts of the geneticists – which are based on genetic theory and studbook history – by analysing the DNA from the reintroduced animals, to quantify the relationships between them, and to ensure that management for out-breeding is on course. Their studies are at present confined to the herd at Taif in Saudi Arabia, where the animals are regularly caught for vaccination against TB (which afflicted their parent herd at Thumamah, near Riyadh), and bled. Without blood samples (or other tissues) the analyses cannot be done.

The Arabian oryx programme, then, has been and continues to be a heroic effort; an example of what can be achieved. Yet, as all informed protagonists emphasise, there is no room for complacency. Without monitoring and management, things could still go wrong.

Finally, there are stirrings of another recovery and reintroduction programme that over the next few decades may re-enact that of the Arabian oryx. Late in 1990 the King Khalid Wildlife Research Centre at Thumamah received two pairs of Saudi dorcas gazelles from the Al Areen Wildlife Sanctuary in Bahrain. The King Khalid Wildlife Research Centre is run by the Zoological Society of London, one of whose scientists is Doug Williamson; and he observes in ZSL's *Lifewatch* magazine that the Saudi dorcas 'is almost certainly extinct in the wild and is known to exist only in one small captive herd of 27 animals' – though there may be others in private collections in Saudi Arabia. The plan now is to organise their breeding, and surely in the foreseeable future to reintroduce them in viable numbers to some safe patch of wilderness.

Yet we must ask to what extent the Arabian oryx reintroduction can truly be seen as a precedent for other reintroductions. Similar plans to reestablish scimitar-horned oryx and addax in West Africa have not yet overcome the diplomatic problems. It has proved remarkably difficult to find a suitable home for Przewalski's horse in all of Mongolia (though we may hope that the initial problems are now solved). Mark Stanley Price emphasises, too, that the biological problems raised by the Arabian oryx are far less daunting than those presented by many other animals. Primates, reintroduced to tropical forest, present the sharpest contrast.

Deserts easy, forests difficult

Mark Stanley Price has probably done more than anyone to ensure the success of the Arabian oryx programme. But he is the first to dampen euphoria; to point out, first, that the Arabian oryx is not yet out of the wood (at least not quite); and secondly, that other introductions may well prove much more difficult. Three general sets of conditions have to be met: diplomatic/logistical; social; and biological/behavioural. Arabian oryx seem especially favoured in all three areas.

On the diplomatic/logistical front, we may note that comparable plans to introduce or reintroduce the scimitar-horned oryx and addax into various sites in Africa have not yet got off the ground, despite lengthy negotiations. We should not give up; but the fact is that not every country has the political structure that can enable big decisions with far-reaching social consequences to be taken quickly, as is often the case in Arab countries. In addition, desert land in general is cheap, and large areas of the Middle East are thinly populated, and although there are very few places that escape human influence these days, there is suitable territory for Arabian oryx. It has proved remarkably difficult, by contrast, to find a suitable site to release Przewalski's horse in Mongolia; though such a release is on schedule for late summer of 1991.

On the social front, we may note simply that oryx are easy animals to get along with. Oryxes eat desert grasses and shrubs; they do not raid crops, as elephants or pigs or chimps may do. They do not eat livestock or people. No one to my knowledge has ever been gored by a wild oryx; if they have, they must have tried very hard to provoke it. It is hard to imagine, indeed, that anyone has ever even been frightened by a wild oryx.

Carnivores, however, whether mammals, birds, or reptiles, provoke all kinds of fears. Sometimes these are to some extent justified. Tigers still eat humans in India (and cattle and goats). Many wild carnivores kill livestock if they get the chance: cheetahs in sheep-pens behave like foxes in chicken-runs, and may kill an entire flock in an evening. Worse, wild carnivores are *perceived* to be dangerous, even though (as I will discuss futher in chapter 7) a wild wolf (say) is far less dangerous to human beings than many a big, fierce, and ill-trained dog that you may encounter any day in a public park. Thus, programmes to reintroduce carnivores are bound to meet local opposition, which must be allayed if the endeavour is to succeed. Yet this is possible. As we will see later, red wolves are now being reintroduced into the Carolinas, in the ultra-civilised eastern United States.

The biological points are of two main kinds. First, some environments simply are much easier for an animal to cope with than others. Secondly, one of the revelations of the past 20 years has been that animals need to *learn* a great deal more than biologists ever supposed, if they are to survive. They are not simply born with all the 'instincts' that they require for survival already hard-wired into their nervous systems. They have to learn, or at least to practice and perfect, the general skills of their species: hunting, foraging, nest building, avoidance of predators, and so on. They have to learn many of their social skills: courtship, childbearing, who to defer to and who to dominate. They also have to learn the *ad hoc* requirements of the particular environment they find themselves in – what Hartmund Jungius of IUCN has called their 'traditions'. Hill sheep farmers have long been aware of this: they like to leave a few old and possibly barren ewes in the flock at the end of each season to guide the next generation through the intricacies and hazards. Again, I will discuss all these behavioural issues at greater length in chapter 7. Suffice to say here that the proportion of skills that must be learnt to the proportion that is hard-wired differs from species to species; but every animal must at least learn the necessary local traditions (and must develop or be born with the skills that are necessary for learning – such as memory, ability to make mental maps, and so on).

Again, the Arabian oryx seems favoured. For animals that do not habitually live in deserts, deserts are hideously inhospitable. For animals that do live in deserts, they are easy. They are two-dimensional; the animals cannot leave the ground. The oryx's ability to find its way around is miraculous; yet as we have seen, it quickly commands huge areas by pursuing the relatively simple strategy of following its nose. There are few predators (effectively none, save human beings and the very occasional raven, in the modern Middle East). The diet is simple, for animals equipped to cope with it; most of the native plants are edible, if you can crunch and digest prickles. The oryx released into Oman did have to learn social skills, and the traditions of the place; yet again it seems that much of their survival behaviour is hard-wired. They sought shade instinctively. They did not need to be taught the technique for finding new pastures.

Contrast all these advantages with life in a tropical forest. The canopy is archetypally three-dimensional, and enormously complex; there may be thousands of potential routes from one place to another. It is immensely variable; no two branches are the same (and they may change from day to day). Predators abound, of many kinds. There are thousands of species of plant, but most are toxic. The ones that are edible are numerous, and yet in the minority; and, of course, all are patchily distributed and most are available only at certain times of year. Much of the food is hidden, under bark or tucked among the leaves of epiphytic plants like the ubiquitous bromeliads (which are related to pineapples but include – among the epiphytic species – the famous 'Spanish moss' that features in so many novels from the US Deep South).

For all these reasons, then, the reintroduction of primates to their various patches of forest worldwide is beset with difficulties that devotees of oryx hardly encounter. Furthermore, primates are great learners – as we should realise, being primates ourselves; and if they are raised in captivity and put back into an environment without friendly natives to help them along, they may find it very hard indeed to learn what they need to know. None the less, there have been and are many attempts worldwide to reintroduce primates to the wild: both monkeys and apes. The difficulties have been horrendous, and they continue. But preparation of primates for the wild, and their management and monitoring after release, is now a major and growing field of conservational endeavour. I will discuss the general approach to the training of primates for the wild in chapter 7. In general (though with many more complexities) the training follows the pattern we have seen in the case of the oryx, of a gradual increase in complexity, from the cosseted cage, through enclosure, to protected wilderness, to wilderness that is as wild as anywhere can be these days.

But it is appropriate to look at two primate reintroduction schemes at this point. The first of these ranks with the Arabian oryx as a modern classic: the golden lion tamarin.

The golden lion tamarin

Golden lion tamarins are small, beautiful, hairy, and literally golden monkeys that share the family Callitrichidae with the marmosets. The Callitrichidae (like the Cebidae, which include the woolly, spider, squirrel, capuchin, titi, and owl monkeys, alias douroucoulis) are all South American. In historical times the golden lion tamarins lived throughout the eastern coastal forest of Brazil around Rio de Janeiro; but today, only 2 per cent of this forest remains, the rest having been logged, chopped for firewood, burned, or simply swept aside to make way for ranches and (recently) weekend cottages and condominiums. By the 1970s there were probably only a few hundred golden lion tamarins left in the wild (in small, largely scattered, and individually non-viable populations) and less than 100 in captivity.

But in the late 1960s and early 1970s, the fortunes of the golden lion tamarin began – one hopes – to be restored. In Brazil, in 1974, the tireless efforts of Adelmar Coimbra-Filho, who has toiled heroically for Brazil's wildlife, were rewarded by the creation of Poco das Antas Biological Reserve: with 5,000 hectares largely devoted to golden lion tamarins. In the late 1960s the now defunct Wild Animal Propagation Trust convened a conference on 'Saving the Lion Marmoset' (the names of animals can be somewhat flexible). In 1973 the first international studbook for golden lion tamarins was begun; it was taken over (together with species coordination) in 1974 by Devra Kleiman of the National Zoological Park (NZP) of the Smithsonian Institution Washington, DC. The cooperation that has grown up since then, between Brazilian authorities and biologists, and conservationists elsewhere; the quality and range of the supporting studies and campaigns; and the interplay between activities in the wild and among captive breeders have truly made this enterprise a benchmark for all conservationists.

As part of the overall campaign, NZP began a research programme in the early 1970s. At that time, there were fewer than 80 animals in captivity, and they seemed doomed. They did not breed well, and deaths exceeded births. Outstanding among a wide spectrum of findings was that golden lion tamarins breed most rapidly in monogamous pairs. To be sure, they normally live in slightly extended family groups; but when they do, all except the dominant pair are reproductively suppressed. A similar phenomenon (though not necessarily produced by the same mechanisms) is seen in mammals as diverse as wolves and naked mole-rats (subterranean rodents from arid East Africa) – and of course (to an even greater degree) among the extreme social insects such as termites, bees, and ants. However, juvenile golden lion tamarins help to look after the youngest animals, and by doing so learn skills that are demonstrably essential, for their later social, sexual, and reproductive development. Finally, female golden lion tamarins are – as Devra Kleiman says – 'incredibly' aggressive towards one another. In general, once all these behavioural and social niceties were appreciated, and curators began to keep their golden lion tamarins in appropriate groups, their reproduction began to go into positive balance.

A small snag was discovered in the late 1970s. An animal at NZP was born with a diaphragmatic hernia, and later it seems that 5–10 per cent of the captive animals were affected. At first it was suspected that the condition might be a simple genetic disorder, of the kind that should be removed by selective breeding (as we have seen is practised among Przewalski's horse). Now, however, it seems that the character is polygenic, or perhaps is not genetic at all. In any case, it cannot reasonably be eliminated by breeding, and it seems simply to be a fact of life, to be dealt with *ad hoc*.

But the population grew. By the late 1970s, NZP began to send golden lion tamarins to other breeding centres (though of course no money changed hands, and all the animals remained an official part of the parent population). By the mid 1980s, the captive population was growing by 20–25 per cent per year; and by 1983 there were about 370 animals in captivity. The problem from then on has been to *reduce* the rate of population growth, while maintaining genetic diversity: the inevitable clash between demographic reality and genetic variability that we discussed in chapter 4. The International Studbook now records around 500 captive golden lion tamarins, in about 80 institutions.

But the wild population, without the benefit of protection and managed breeding, remained precarious. A survey in 1980 suggested that Poco das Antas contained only about 100 individuals, and that there were only tiny pockets outside the reserve. Much more recent studies suggest that the wild population had less genetic diversity than the captive population; presumably because the wild population originated with founders that may have come from various parts of the tamarin's range, while the Poco das Antas animals were all from one place. In any case, the wild population seemed unviable. Thus it was that at the beginning of the 1980s Devra Kleiman discussed formal collaboration between the captive breeding efforts, centred on NZP, and the Brazilian authorities, with Adelmar Coimbra-Filho and WWF.

The campaign that has developed from this collaboration has five components: management of and research into the captive population; field research into the wild population; protection, management, preservation, and restoration of the habitat; public education – not least to ensure that people take care of the tamarins outside the Poco das Antas; and reintroduction. Note that all five are essential. As things are (and regrettably) habitat protection alone could probably not have saved the golden lion tamarin from extinction – although it probably would have done, if it could have been initiated on a larger scale, some time earlier. In the mid-1980s less than 40 per cent of Poco das Antas itself was still forested, and only 10 per cent was pristine forest; part of the task, then, has been to try to re-establish the forest, which *inter alia* has required liming of the ground to tip the balance back from grasses to trees.

Note too that reintroduction was not perceived as the panacea, but as part of a larger plan. As things are, though, it was an essential part of the plan; for without a steady trickle of genes from the captive population, the wild population seemed doomed to enter the spiral of extinction that begins with and is driven by inbreeding. Note, too, that the context of the golden lion tamarin reintroductions has been quite different from those of the Arabian oryx. This is not an attempt to re-establish an animal that no longer existed in the wild. This, rather, is an attempt to bolster a wild population that is ailing.

Devra Kleiman describes the early reintroductions in *Primates: The Road to Self-Sustaining Populations*.[17] Fifteen captive-born animals were sent to Brazil in November 1983, though four subsequently died in the six-month quarantine. The remainder were moved to Poco das Antas in May and June of 1984, where they were first held in large cages to acclimatise them. Fourteen animals were released between May and July (some had meanwhile been born!) including one family group of eight, and three adult pairs.

They did not do well. By June 1985, 11 were dead, or had been rescued. Causes of death included exposure, disease, starvation, snakebite, predation and disappearance. Disease was the biggest single cause of death – what was probably a virus caused diarrhoea and dehydration in five animals in a family group in February 1985. This was a bolt from the blue, but other deaths were more salutary. For example, the captive-bred tamarins, which should have been marvellously agile little monkeys, were not good climbers, and particularly disliked the difficult passage from tree to tree, along thin twigs. They tended to move instead along the ground. It was thus that one fell foul of a snake, and another was caught by a feral dog. The contrast with the newly-released Omani oryx was striking – among whom, as we have seen, casualties were nugatory.

Some critics – those who reflexly oppose captive breeding – have taken this initial toll as evidence that captive breeding and reintroduction are doomed. The real point, however, is simply that they can be extremely difficult. Since those early days, as we have said (early, though only six years ago at the time of writing) reintroduction has invariably been preceded by training, in climbing, feeding, and all other skills, which I will discuss in chapter 7; training that in fact implies more naturalistic husbandry throughout the animals' lives. With such preparation, it now seems that reintroductions even of the most difficult species can succeed. Sometimes reintroduction may take more than one generation; cage-reared animals should adjust to a reserve, and their offspring, born in the reserve, can adapt to wilder country, and so on. This has been found at Apenheul in Holland, where primates run more or less free, as again will be discussed in chapter 7. We must accept, though, that the wilderness is a dangerous place. Mortality is high even among animals that have always lived there. Only among captive populations has it become habitual to *expect* the majority to survive.

Today, the tamarin conservation project continues, much as it was first envisaged. Reintroductions continue, but in a more structured way than originally. The captive population seems established. The species seems safe from extinction, at least for the time being. As with the Arabian oryx, we can allow ourselves a little cautious optimism. Among many other primate reintroductions, one that strikes me particularly (because I knew the animals concerned some time before their reintroduction was planned) involves woolly monkeys. In this instance, remarkably, captive-bred animals are being employed to lead wild-born animals, and protect them in the wild.

Return of two woollies: captive-born tutors

The story of this particular reintroduction again begins in two places: at Noah's Park, in Brazil; and at the Monkey Sanctuary in Cornwall, in south-west England. To begin in Brazil: Noah's Park is way up the Amazon, 30 miles from the fantastic city of Manaus, which was built for rich rubber merchants and is complete with opera house. The Park was established first as a farm to derive protein from capybaryas (the world's largest rodents), but this proved uneconomic. A Dutch biologist, Marc van Roosmalen, is now seeking to re-establish the park as an economic going concern and a source of local income – not as a ranch for novel livestock, but as a centre of 'eco-tourism'. To this end, he is seeking to restock it with local animals, including woolly monkeys.

Woolly monkeys are endangered, but local rubber tappers kill the adults for meat none the less, and sell the infants in the local market as pets. The trade is illegal; but one baby woolly is worth a month's wages. Once the babies have reached the market, intervention has seemed pointless, because if they were confiscated at that stage, there would be nowhere to put them. One of the functions of Noah's Park, however, is to provide these orphans with sanctuary. In practice, then, Marc van Roosmalen has been taking the very young confiscated monkeys back to his home, where he and his wife ensure that they are healthy. Then he takes them back to Noah's Park, where they remain in enclosures to habituate them, just as is done with golden lion tamarins (or indeed oryx or any other animal). Then they are released, though they are provided with supplementary feed, firstly to give them a home base and ensure they do not wander too far too soon, and secondly to supplement their own first efforts to forage for themselves. In fact, supplementary feeding is a permanent feature, for it cannot be certain that the 30,000 hectares (100 square miles) of Noah's Park will provide all the food they need, on every day of the year.

The method works: Marc van Roosmalen had made many successful reintroductions. But there is a snag. Wild woolly monkey troops do not generally contain more than about 30 animals, which should include at least one dominant adult male, generally with one or several subsidiary adult males, plus adult females, juveniles, and infants. Without the leadership of the experienced animals, the troop cannot thrive. Van Roosmalen has been able to create one reasonably balanced troop. But most of the animals he brings in from the market are very young: virtually infants. They are far too young to cope by themselves, but the existing troop is too large to absorb any more. Thus, the monkeys he acquired in 1990 (and continues to acquire) could not safely be released, and could not have formed a coherent social group if they had been.

This is where the Monkey Sanctuary comes in. It was founded in Cornwall in 1964 by Leonard Williams, father of the internationally renowned guitarist John Williams. Williams, like van Roosmalen, was concerned first of all to provide sanctuary; for the organ-grinders' monkeys and pets that in the 1960s were still being imported to Britain. He gave over his Cornish house to them, and built enclosures all around that were extended into the house; and, as time passed, established a truly coherent and balanced social group of woolly monkeys. At that time, most zoos who had woollies at all kept them in pairs, and then wondered why the males – who without a troop to look after were deeply frustrated – took out their aggression on the females. Woolly monkeys in the zoos of those days rarely bred.

Leonard Williams's monkeys did breed, however. Now they are in their fourth generation. Williams himself died in 1987, and the head keeper since then has been Rachel Hevesi. But she now finds herself with a problem that is effectively the opposite of the one that taxes Marc van Roosmalen. The Monkey Sanctuary colony now numbers around 20 animals, which is about as many as it can comfortably take; and is almost as big as a woolly monkey troop likes to be. It is also well balanced socially, with two adult males (Charlie, the boss, and his uncle, Django). But this balance will not last much longer, for it also has three 4-year-old sub-adult males who were born at the Sanctuary – Ricky, Ivan, and Nick. Those three will soon be wanting troops of their own to lead. Already they show signs of restlessness.

Rachel Hevesi saw it as part of her task when she first took over the Sanctuary to make contact with all other woolly monkey centres – not least to ensure that the necessary outbreeding should take place. Thus she learned of Marc van Roosmalen, and visited him at Noah's Park in 1990. There they hatched the plot to solve both their problems. Two of the Sanctuary sub-adults – probably Ricky and Ivan, for genetic reasons – should be taken to Noah's Park and serve as leaders for the orphaned infants who were too young to face the world on their own. This is a remarkable development: captive-born animals serving as leaders, in the wild, of animals who themselves are wild-born.

Yet there is every reason to hope and expect that the scheme will work. For one thing, and essentially, Ricky and Ivan have already shown, in Cornwall, that they like to look after younger animals, in the way that many monkey species do. In Cornwall they do not live free: but they live in naturalistic enclosures, with natural vegetation (so they know how to climb and find their way round); and they know how to forage for food. Noah's Park is 'wild', but not unprotected or unrestricted; it is confined by a river on one side, a fence on another, and roads on two sides, and contains very few predators. It is anticipated that the wild-born youngsters – though they were very young when captured – will be able in their turn to introduce Ricky and Ivan to the wild vegetation.

Also encouraging, in various ways, is the behaviour of the group that is already established in Noah's Park. During Rachel Hevesi's first visit to the Park, an infant female was brought in who was only 10 weeks old. Her survival was a miracle. Rachel Hevesi bottle-fed the baby, but feared that it would become imprinted on her, and grow up as a neurotic pet. She determined, then, to encourage the troop in the reserve to foster it. So she went in to the park with the baby and called to the monkeys (using the chirrup call she had learnt from the woollies at the Sanctuary). They were very curious. The urge to look after babies is very strong with them. One young female in particular – though she was one of the monkeys who normally stayed well away from humans – could not resist. She came down to see the baby. To her surprise (and the baby's) Rachel placed the baby on her back.

Within two weeks, extending the contact each day, Rachel was able to foster the baby entirely on to the troop (despite its initial howls of protest). The monkeys could not feed the baby; but they recognised Rachel as its rightful mother, and brought it back to her to be bottle-fed. After Rachel returned to England, they brought it to Marc instead. Now the young one is no longer a baby. She is a fully integrated member of the troop and – pleasingly – she is one of the individuals who is most stand-offish towards humans.

Nicky and Ivan are due to travel to Noah's Park, with Rachel, late in March 1991; almost exactly at the time of writing. There is every reason to expect success. But even if some accident upsets this particular venture, Leonard Williams, Rachel Hevesi, and Marc van Roosmalen have abundantly demonstrated that when intelligent animals are treated humanely and with empathy, their capabilities can exceed all imaginings.

I will return to these behavioural aspects, and say more about the Monkey Sanctuary, in chapter 7. Here, I will discuss reintroduction problems of different kinds: those raised by carnivores; and the special problems and advantages of birds.

The red wolf

The story of the red wolf of the south-eastern United States – how it was snatched from the wild when apparently doomed to extinction; bred in captivity; and then reintroduced into the Carolinas, beginning in 1987 – in many ways illustrates the general problems that beset the conservation of wild carnivores. Yet in one way it is unique. For although many carnivore reintroductions have now been attempted – from big cats in

Africa and Asia to the swift fox of Canada and otters in Great Britain –
this is the only occasion on which a wild carnivore that has become
extinct in the wild has been reintroduced to a part of its former range. The
case of the red wolf may not remain unique for much longer, however,
because the reintroduction of black-footed ferret, probably to Wyoming,
is planned for autumn 1991. These two stories, then, are comparable with
that of the Arabian oryx.

The red wolf, *Canis rufus*, once ranged from the Atlantic Ocean to
central Texas, and from the Gulf of Mexico to central Missouri and
central Illinois. To the west was coyote territory, and to the north was the
gray wolf, but although the red can hybridise with either (all members of
the genus *Canis* in general hybridise happily) it is considered distinct.
Some say, indeed, that the red wolf belongs to the 'primitive' wolf stock
of North America of one million years ago. But people in general have
not behaved well towards wolves, and by the early 1960s the red wolf was
clearly in danger in the wild.

A recovery programme in the late 1960s was limited in scope; but the
Endangered Species Act of 1973 concentrated conservation efforts, and
the US Fish and Wildlife Service (FWS) collaborated with Point Defiance
Zoo in Tacoma, Washington to begin the Red Wolf Captive Breeding
Program. By late 1975 the FWS declared that it was no longer feasible to
preserve the red wolf in its steadily decreasing range in Texas and
Louisiana, and as many as possible of those that remained were rounded
up for captive breeding. The species was declared 'biologically extinct' in
1980. In 1984, AAZPA established a Species Survival Plan.

Port Defiance Zoo set up 12 pens, 30 metres by 30 metres. These were
out of view of the public – 'off exhibit'; proof (if proof were needed) that
zoos do not exist solely to entertain people and indeed do some of their
most interesting work out of sight. The wolves bred well, and were
distributed among other zoos. From the start, the intention has been to
reintroduce them to the wild. Efforts have been made to ensure that the
captive animals do not become too trusting of humans – partly because
wild carnivores that are habituated to humans are in general more
dangerous (or at least more alarming) than those that keep their distance.
Mistrust is good for the animal, too, for animals that are too trusting are
easily killed. The captive wolves are also in general kept within large
outdoor pens, where they can continue to hunt whatever wanders in, and
where natural selection can take its toll. There are shades here of the early
husbandry of captive Père David's deer.

The build-up to the final release, in 1987, contains many lessons. One is
that it is doubly important when reintroducing carnivores to gain public
support – especially for an animal that has been driven to extinction
largely through public antipathy. Many public meetings were held.

Intriguing points came to the fore. Deer hunters, predictably, feared for their intended quarry, while welfarists feared for the safety of the wolves. In his discussion of the red wolf at the London Zoo symposium on reintroduction in 1989, Donald Moore of Burnet Park Zoo at Syracuse, New York singled out three significant reasons for the eventual success of the pre-introduction negotiations.[18] One was the assurance that people who inadvertently ran over a wolf would not be prosecuted – provided it was an accident – as the terms of the Endangered Species Act otherwise seemed to allow. Another was that reintroduced animals should be given 'experimental population status', which means that their movements can be closely monitored with radio collars, that any that stray outside their reserve can be returned, and that individuals perceived to be dangerous can be taken back into captivity. Black-footed ferrets may also have to be given experimental population status, to enable the human population around them to conduct their own lives more freely, if their reintro-duction gets under way. Finally the necessary diplomacy succeeded, said Donald Moore, largely because the FWS personnel who argued the case were extremely amiable, and became good friends with the local people.

Thus, in 1986, when the captive red wolf population stood at around 80, eight individuals – all young adults at 4–6 years, and second-generation in captivity – were selected for reintroduction. Their diet was switched from standard zoo rations to road-kills and live prey, and they were prepared for 'slow release' (that is, stage-by-stage). The first two – the historic first two – were released on 14 September 1987, into the Alligator River National Wildlife Refuge (ARNWR) in North Carolina.

By the end of 1989, 39 red wolves had been released into the ARNWR and on to Bull's Island in South Carolina. Twelve had died, including three hit by cars; though the rest died from natural causes, including two eaten by alligators. By 1990, the population among zoos and refuges in the United States stood at around 125. This looks like a success.

SOME BIRD STORIES

The conservation of birds is in some ways easier than that of mammals, and in some ways more difficult. One general problem for captive breeders of birds is that in many species the males look exactly the same as the females. In many, the sex can be definitively determined only by laparoscopy: opening the abdomen and looking at the gonads, which are internal. Blood tests to look at the sex chromosomes in the blood cells are often unhelpful because many birds have many small chromosomes, which are very difficult to tell apart. Scientists at London's Institute of Zoology are now exploring molecular methods: looking for DNA differences between the sexes. But in any case sexing is a job for experts, and at present involves catching and probably performing minor surgery, which for wild birds (albeit captive wild birds) is traumatic. Many a zoo in the past had many a breeding failure as two males or two females sometimes cohabited for decades.

Eggs, in general, are on the side of the conservator. Many kinds of bird will lay more eggs if the first-laid are removed; this being known as 'double clutching'. Some birds, like the once-threatened bearded vulture of the Alps and Pyrenees, lay two eggs but then the one that hatches first kills the younger; a practice sometimes known as 'Cainism'. So if the first egg is removed, the young can be raised separately. As we will see in chapter 6, reproductive biologists are now going to enormous lengths to try to persuade endangered but non-fecund mammals to produce more offspring than they normally would. But the output of many birds can be raised prodigiously by a little sleight of hand. Thus was the population of North American wood-duck brought from the brink of alleged extinction at the turn of the last century, to a level that now allows a 'cull' of 750,000 per year. The main theoretical snag is that some birds at least become 'imprinted' on the keepers who hand-rear them, and refuse henceforth to live normal lives. Yet there are ways round this too. San Diego keepers employed condor-head glove-puppets to teach hand-raised California condors that they are, indeed, condors. Bean geese in Scandinavia have been fostered under Canada geese; species similar enough to make little difference to their psyche. Success is not guaranteed, however; for although whooping crane populations in the United States have been increased by fostering under sandhill cranes, the fostered whooping cranes have not bred convincingly themselves, which at least suggests that their behavioural skills are lacking. At least, though, the sheer raising of numbers is generally easier in birds than in mammals; and any other problems that arise must be solved *ad hoc*.

Eggs are easy to transport, too; which in general should favour attempts at reintroduction. Indeed, reintroduction has often been attempted – and introduction of species to habitats they never formerly occupied has been attempted even more often. In fact, according to speakers at an international symposium at the Wildfowl and Wetlands Trust headquarters at Slimbridge in Gloucestershire in 1988, serious attempts at reintroduction and introduction now total at least 1,670. Most of the introduced (as opposed to reintroduced) species have been wildfowl and game birds, which people like to look at and to shoot. Thus the British landscape is adorned by such beauties as the mandarin duck, Carolina duck, and Canada goose. Perhaps we should not be too purist at this late stage; but as we saw in chapter 2, introduction of aliens is a major cause of extinctions among native animals. Reintroduction and introduction are, in general, very different undertakings, with very different motivations.

Flight may both facilitate and complicate attempts at reintroduction. Birds escape predators more easily and generally more instinctively than the land-bound golden lion tamarins that were released before the need for training was appreciated. But an early attempt in the 1980s to release captive-bred pink pigeons into the botanical gardens at Pamplemousses in their native Mauritius came unstuck because a high wind blew the birds off course. Flying vultures (like the reintroduced bearded vultures of the Alps) are easy to monitor, which is an essential ingredient of re-introduction. But it is also in general more difficult to confine flying birds than mammals in safe reserves; and birds of prey in particular that fly far afield are particularly prone to poisoning on hostile farms.

So there have indeed been many serious introductions, reintro-ductions, and translocations. But the success has been erratic – a mere 15 per cent in the United States, according to one speaker at the Slimbridge Symposium. To illustrate the range of problems and possibilities, I will discuss just five examples. One – the bearded vulture – seems so far to be a success. Rothschild's mynah and the California condor are about a third of the way into the captive breeding reintroduction cycle; it is far too early to run up the flags but there are reasonable grounds for optimism for these two creatures which, again, have been snatched from the grave. Parrots as a group could undoubtedly benefit from captive breeding (and to some extent are benefiting); but between them they illustrate another swatch of difficulties among which human follies are prominent. Finally, just to finish on a note of good biology and high motive, the Hawaiian goose or ne-ne has undoubtedly been saved as a species, which is in itself a triumph. But although it has often been presented as the textbook success, attempts to reintroduce it cannot yet be said to have succeeded.

The bearded vulture

The bearded vulture or lammergeier is a magnificent bird, with wings that span three metres and carry it high above the Alps and Pyrenees. Once the bird was almost common. But it is the fate of birds of prey to be seen as a threat; small species, like the red kites now being reintroduced to England, were seen as a threat to game-birds; and big species, like the lammergeier, to lambs, chamois, and even children. The facts are rarely allowed to enter the case: the fact in this case that the lammergeier is a carrion feeder, whose speciality is the breaking of bones, which it dashes against rocks at the nadir of a long, falling swoop. But the locals chose none the less to hunt the lammergeier almost to extinction.

The programme of salvation is organised by WWF in Austria and is based at the Austrian Alpine Zoo in Innsbruck. Multiplication was enhanced as we have seen, by removing the younger egg that would otherwise be killed; but the youngster can be raised by its parents if it is returned to the nest when big enough to fend for itself – or perhaps may simply be partitioned off from its overweening sibling, though still visible to its parents.

About three weeks before they are ready for independent flight, the young birds are taken to a sheltered ledge, high in the Alps in Salzburg province, about 60 kilometres from Salzburg. There they learn to fly and forage. Mark Stanley Price has commented that animals that are easy to eliminate – perhaps because they are easy to see, like the Arabian oryx and lammergeier; perhaps because they are easy to poison, like birds of prey in general – can also be easier in some ways to reintroduce. Nowadays, then, the lammergeiers are fed not with poisoned meat, to finish them off, but with wholesome carcasses to keep them going. I have not been privileged to see the lammergeiers being fed, but have watched Cape vultures with their dove-grey plumage descending to feed on ancient mules, laid out for them in their 'vulture restaurant' at De Wildt near Pretoria, better known for its cheetahs. The lammergeiers in the Alps are also helped by the local avalanches, which entrap entire herds of chamois and release them slowly through the thaw, as if from a deep-freeze.

By March 1989 13 birds had been released of whom, in 1990, 11 (and perhaps 12) were surviving. The programme is planned to continue until the first decade of the 21st century, when the birds should be re-established as a viable population. Some of the released birds have travelled as far as Italy and Switzerland. Though born in captivity, and without instruction, they have retained their instinct for breaking bones. It does not seem to be an easy way to make a living. Sometimes a bone needs 30 swoops to break it.

The lammergeier was doing well before people hunted it; and now they have been persuaded to stop hunting it (and numbers have been boosted by breeding) it seems set fair for recovery, for its havens in the Alps still exist. The ecology of the California condor was far more precarious, and the politics of its rescue more complex.

The California condor

The California condor (*Gymnogyps californianus*) and the Andean (*Vultur gryphus*) are the world's two remaining condors, within the New World family of vultures, the Carthartidae. They are much the same kind of bird; yet they are different enough to be placed in separate genera. The Andean is doing reasonably well both in the wild and in captivity but the California is now extinct in the wild, and captive breeding is its final hope. The California was probably at its most successful in the late Pleistocene period – beginning a few hundred thousand years ago, to around 10,000 years ago – when it apparently fed on the carcasses of the elephants, giant ground-sloths and rhinoceroses that then abounded in North America. Indeed there was food enough for two North American condors – the other being the long extinct *Breagyps clarki*, which had a long beak that probably allowed it to feed on viscera deep inside big carcasses, as the Griffon vultures of Africa do today. But the big animals disappeared, probably because they were slaughtered by human beings. The mystery, then, is not why *Breagyps* disappeared, but how it is that *Gymnogyps californianus* managed to hang on, albeit only along the west coast. Some hypothesise that it fed upon the corpses of beached whales – though that does not explain why *Gymnogyps* declined along the coast of Florida. Neither is it clear why *Gymnogyps* failed to occupy midwestern America, feeding on the carcasses of bison.

For whatever reason, the California condor did manage to survive in the west, and indeed began to flourish again as Europeans introduced domestic cattle, whose corpses have become its main fare. But it declined again. By the early 1930s only 100 were left; by the late 1940s only about 60; by 1970, fewer than 50; and by 1980, 25 to 30 at most. They are long-lived, so the decline is even more precipitate than it seems. By March 1986 the wild population was down to five birds, and the one egg that they produced between them had a paper-thin shell of the kind brought about by poisoning with DDT; and indeed the egg contained high concentrations of DDE, which is a metabolite of DDT. By the autumn of 1986, the wild population was down to three. They – at last – were taken into captivity.

Everything has been against the California condor throughout this century. It has been protected by law since the 1900s, yet shooting continued until the 1980s. It also suffers from loss of feed, for modern farmers do not leave so many carcasses lying about as those of the past. It sits at the top of its food chain, and accumulates anything nasty from the succession of creatures below it. DDT is one of those, as we have seen. Many of the carcasses the condors fed upon in recent decades had themselves been shot, and contained lead. Thus, one of the last California condors left in the wild in the 1980s was clearly ailing, and was finally trapped and brought into San Diego Wildlife Park for treatment (though with much opposition from some local conservationists). The best efforts, 24 hours a day for 15 days, failed to save her. X-ray showed that she had been shot, and had eight pellets in her flesh. But she had died from a ninth pellet that she had ingested. Lead, like DDT, is a cumulative poison.

Clearly, then, by the 1980s the California condor was doomed in the wild. The very few that remained lived a life of extreme hazard; like a miner's canary, one might say, exposing the general precariousness of vast and beautiful California. But the bird did survive in captivity. Los Angeles Zoo had 12 in October 1986, and the San Diego Wildlife Park also had 12.

The only sensible course, it seemed (as indeed it had seemed for some decades) was to capture the last remaining birds from the wild, and build a new population with as many founders as possible, as was done for the Arabian oryx, black-footed ferret, and as far as possible for the red wolf. There was powerful opposition, however. The National Audubon Society (NAS), which is affiliated to the Smithsonian Institute, argued for 'protection in the wild'. Only by a miracle – literal divine intervention – could three birds in an uncontrollable but largely hostile environment have brought themselves back from the brink. Captive breeding with a broader genetic base did offer hope. Better than captive survival, however, felt the NAS (and the Friends of the Earth and the Sierra Club) was for the birds to 'die with dignity'.[19]

Today, the California condor is gone from the wild, but as William Toone of San Diego Wild Animal Park and Arthur Risser of San Diego Zoo wrote in the *International Zoo News* of 1988 'there is every hope that the captive breeding programme will be successful.'[20] To be sure, captive California condors have not often bred in the past: but 'except for the females kept at the National Zoo [Washington], there appears to have been no attempt to breed them.' Cathartid vultures in general have bred well in captivity, and Andean condors at San Diego have been multiplied by the technique we have noted – removing eggs and encouraging the birds to nest two or three times in the season. Left to themselves, the two condors are slow breeders; they do not breed until they are at least 6 or 7, they lay only one egg at a time, and look after the young until it is in its second year. So one egg every two years is the normal strike rate.

In practice, the California condor does not take badly to captivity. As with the lion, its ostensible dignity is linked to incorrigible idleness. If the birds are ever released into the wild, it may be to the Grand Canyon (which was once part of their range); or perhaps simply into the more confined area of the Sespe Condor Sanctuary in California, where they will be encouraged to stay with feasts of (uncontaminated) carcasses. One plan is first to release two female Andean condors to see if the terrain is suitable for condors in general. In the long term, it is at least conceivable that larger areas of California might become less hostile; and the successors of the 'Die with Dignity' school might have cause to be grateful for the more positive approach.

Rothschild's mynah

Rothschild's mynah, alias the Bali starling, is a beautiful snowy-white bird with sky-blue rings around its eyes and a crested head. It once abounded in Bali, where it is the only endemic vertebrate (that is, the only vertebrate species found on Bali but nowhere else). But in 1984 a survey by the International Survey for Bird Preservation revealed only 200 individuals, all in Bali Barat, which is now a national park but in those days was not. ICPB then joined with AAZPA, the Jersey Wildlife Preservation Trust, and the Government of Indonesia to breed the birds in captivity and restore them to the wild.

Rothschild's mynah probably declined in the wild mostly because of loss of nesting sites and perhaps of food; but the depradations of traders may also have played a part, for they are popular aviary birds. As a result, though, there are at least 500 in captivity. Yet they do not breed reliably in captivity, though the reason is not clear. Lack of genetic variation may play a part. Thus, the US population (of more than 400) evidently had only 37 founders, of whom just five have dominated the stock. Although Britain's population of around 100 birds is apparently self-sustaining, breeding depends heavily on just a few successful pairs.

Britain's endeavours to increase numbers while conserving genes, organised in the late 1980s by Dr Georgina Mace of London's Institute of Zoology, showed just how complicated such schemes can be. Among the traffic planned for 1988, for example, Bristol Zoo was invited to send male number 382 to Harewood Bird Garden in Yorkshire, while taking a male from Harewood (number 256) to mate with female number 381; and also to send number 636 to London for sexing. Edinburgh Zoo was invited to send its male from pair 3 to Lotherton Bird Park in Leeds, and the female to Southport Zoo, while receiving a pair from London once the birds had been sexed. And so the plan unfolds: with 14 institutes (in Britain alone!) to-ing and fro-ing specific birds between them. As we have commented, without the computer, such orchestrated efforts would be quite unmanageable.

Despite the problems of breeding, however, the reintroduction to Bali began in 1987, with 42 birds: 38 bred in the US, and two pairs from Jersey. The plan is to release 500 by 1992. The birds stop first at Surabaya Zoo in Indonesia, where they are trained for life in the world before their release in Bali Barat.

Parrots

Parrots, between them, illustrate many principles of conservation; many lessons of captive breeding; and much about the psyche and behaviour of human beings. As a group, parrots are perhaps the most threatened of all creatures: about 70 of the 320 species of parrot, macaw, amazon, cockatoo, cockatiel, lory, lorikeet, conure, parakeet, lovebird and budgerigar are now thought to be endangered. Most species (as is usually the case for animals in general) are threatened mainly by loss of habitat, for most are creatures of tropical forest. But parrots – bright coloured, intelligent, and gregarious: 'primates of the bird world' – are also the most desired of cage birds; and for some the chief threat is trapping and trading. This is true of the wonderful Hyacinthine macaw, the biggest of all the parrots and adorned in powdery blue; and for the lovely Spix's macaw. Both live in Brazil in land that is not threatened, or even is useless for humans; Hyacinthine generally in open woodland, Spix's in swampland further north. Both are specialists feeding upon the nuts of Buriti palms. Both are especially vulnerable partly because they are so conspicuous, and partly because they must fly large distances from food tree to food tree. No bird flies quite in the arrow-straight way of parrots.

Aviculture and the trade that goes with it play an ambivalent role in the lives of parrots. Whatever the cause of their decline in the wild, many species probably cannot now be saved from extinction except with the assistance of captive breeding. Yet many species breed poorly in captivity; and it happens to be the case that various enthusiasts are among the most proficient of all parrot-breeders. These enthusiasts cannot be called 'amateurs', because some of them earn their living from their craft. Neither are they strictly speaking 'professionals', since few have zoological or veterinary qualifications. In chapter 8 I refer to them all collectively as freelancers.

Clearly, in some ways the freelancers are bad. They have encouraged and indeed may now perpetuate the trade that has exacerbated the plight of many wild parrots, and which is the principal threat to a few. On the other hand, some have the very skills and facilities that are now needed to rescue parrots from oblivion. Some, too, are genuine enthusiasts, dedicated to conservation. Some of those enthusiasts undoubtedly produce more young parrots than they have ever taken from the wild; but some of the most successful and prolific propagators have none the less taken animals from the wild, sometimes in contravention of CITES rulings, and this can be perceived as a cardinal sin. On the other hand, some of those who have broken the law argue that they have in many cases rescued the birds that they have imported from a certain death in the wild; and some of those enthusiasts who argue that point have travelled wide and long in the countries from which the parrots come, and they often know more about conditions on the ground than the lawmakers. The whole area is very, very grey.

However, overlying, underpinning, and generally complicating the entire freelance endeavour is money. Among zoos committed to IUCN-inspired captive breeding plans – SSPs, EEPs, and the rest – money does not change hands. Whatever is done with Javan rhinos, for example, all animals will continue to belong to the Indonesian people (and in an enlightened world we would lose the concept of ownership altogether, and think instead of guardianship, as the Australian aboriginals think of land). But between dealers and fanciers of parrots, huge sums are exchanged. Single birds may be valued at tens of thousands of pounds.

Thus does sordidness creep in. Some freelancers who might otherwise join with others and with zoos in properly coordinated breeding plans are reluctant to do so for fear of theft. Many prefer even to keep their addresses secret. The fear is justified. Break-ins and thefts are frequent. In November 1990 (just to take one recent example) thieves stole a scarlet macaw from Whipsnade Wildlife Park, the country home of London zoo. The bird's parents, who had been wild-caught and whose genes were therefore much desired for captive breeding, had been stolen the previous December. Finally, it is even said that some owners deliberately choose not to breed, so as to increase the rarity of their own animal.

I do not presume to know precisely why some of the owners of the 40 Spix's macaws that are thought to remain in captivity refused to take part in a breeding programme planned by the ICPB; but they did. By mid-1990 there was only one Spix's left in the wild, and we can only hope that the ICPB will succeed in its rescue plan for the Spix's with the birds that it can lay hands to. A refuge is planned, with a breeding population based on some at least of the remaining captive birds, in the Spix's own native habitat.

Finally, the creature that is widely perceived to be the classic conservation story among birds: the Hawaiian goose, or ne-ne.

The Hawaiian goose

The saving of the Hawaiian goose (*Branta sandvicensis*) is perceived as one of the classics of modern conservation, to rank alongside the Arabian oryx; and to a large extent that is justified. By 1952, only 32 individuals remained, and they were clearly ailing, besieged not least by introduced mongooses. Sir Peter Scott and his colleagues at the Wildfowl Trust (now the Wildfowl and Wetlands Trust) conceived a plan to remove them from the wild and establish a breeding colony at Slimbridge – and then to re-establish the ne-ne in the wild. Peter Scott was one of the pioneer presenters of natural history on television; and I remember even in those far-off days, when I was still at primary school, thinking what a bold and wonderful plan it was. I still do.

The birds bred very well at Slimbridge, and now, it seems, can be found in zoos and refuges worldwide. There have been plenty to spare for reintroduction. By now, indeed, more than 3,000 have been released. Yet the final phase of the dream has not been realised. Hawaiian geese are not re-established in the wild in a viable population. So what has gone wrong? Mongooses are still a problem. Worse, though, perhaps, is that the biology of the ne-ne has been misconstrued. The last birds in Hawaii lived high on the volcanoes, amid the lava. But this is probably not their preferred home. Palaeontologists in recent years have now discovered many ne-ne bones on the islands of Hawaii, and mostly on lowland sites.[21] They have also found bones of another goose and six ducks, all now extinct; proof here of Jared Diamond's argument, from chapter 2, that the extinctions wrought these past few millenia among island species in particular far exceed what was previously suspected. But the extinct goose was probably flightless, and the extinct ducks certainly were, and Thane Pratt of the Department of Forestry and Wildlife in Hawaii suggests that the Hawaiian goose survived only because it was not flightless, and could take off for the highlands. But the birds do not like being there. The released pairs do not breed every year, and rarely raise a gosling to adult size. It seems there just is not enough food.

However, Hawaiian geese introduced to lowland areas have bred; and Thane Pratt thinks there may well be a case for a formal reintroduction programme to a lowland site – somewhere that predators can be controlled. It certainly seems to make sense. Meanwhile, we can at least rejoice that the first part of Peter Scott's vision has been realised. The Hawaiian goose has been saved from extinction. Though reintroduction has largely failed so far, we can at least try again. If the ne-ne had 'died with dignity', this would not be an option.

It is tempting to turn this chapter into a hobby and continue it forever. There are hundreds of examples to draw upon, each with its own intricacies. But enough has been said to show that captive breeding is a complicated business, and that reintroduction complicates it even more; and to show, too, that no two cases are exactly alike, and indeed that the general terms 'captive breeding' and 'reintroduction' cover a multitude of endeavours. We have shown, too, that no captive breeding and reintroduction scheme is ever 'finished'. Once animals are put back in the wild, the monitoring must continue; and, increasingly and for the indefinite future, we can expect a continuing and orchestrated flow of genes between the captive populations and the wild.

We have repeatedly mentioned the increasing need to enhance the reproduction of endangered animals. The special techniques for achieving this are the theme of the next chapter.

SIX

THE FROZEN ZOO?

It is easy to fantasise about the future of technology. If you leave the time-scale open – if you do not say when you expect events to take place – your predictions can hardly be disproved. After a century that has produced supersonic flight, television, the laser, particle accelerators, the computer, and genetic engineering, almost any technology must seem credible. Provided we do not break what Sir Peter Medawar called 'the bedrock laws of physics' – anything goes.

Thus, I have heard an eminent scientist opine that in some future century, we might recreate organisms at will – animals, plants, bacteria, or whatever bizarre forms we chose to imagine. The method would be simply to synthesise DNA that would code for the particular proteins which, by their interactions, would give rise to an organism of our choice; and then to place that DNA in a milieu (which would probably be some suitably modified cytoplasm from an existing organism) in which it could work its magic.

Any follower of modern biology would immediately perceive a long catalogue of huge difficulties. The greatest is the present dearth of knowledge of all the steps that lie between the raw code of DNA and the stupendous flow of interactions in a finished organism. On the other hand, the scientist who made the point to me is a biotechnologist. He knows the details of the problems at least as well as anybody. But he also appreciates (as all biologists must) how enormously far we have come since the early 1950s, when Francis Crick, James Watson, Maurice Wilkins and Rosalind Franklin first revealed the structure of DNA.

As a molecular biology watcher now I am of course struck by the magnitude and number of problems still to be solved, and at the number that have hardly yet begun to be framed. But I am also struck by the sheer pace of progress; the fact, for example, that there are now entire directories of gene sequences, and in many cases it is known precisely what proteins those genes produce, and how those genes operate. In many cases, too, molecular biologists already know how to cause particular genes to express in particular contexts, and to lie 'dormant' in others. So on all fronts, the research that could enable us one day to build organisms virtually at will is now being carried out. We may yet discover more 'bedrock laws' that will prevent us from the ultimate construction, but there is no reason at present to suppose that we will. If we think ahead 200 years, or 500, as we do throughout this book, then – barring ecological or social collapse – we can allow our imaginations to let rip.

Yet surely such a prospect is hideous – that we can simply create organisms at will, if we wipe out the present ones? We are back to the point of chapter 1: that such a vision as this is in the realms of hubris – a piece of blasphemy; and, indeed, that we do not have the imagination to conceive of creatures one-thousandth-part as wonderful as the ones we have inherited. We may contemplate repair-jobs in the future, if forced into it. We may in some cases seek to eliminate genes that are known to be undesirable – a form of eugenics; as we have seen in chapter 5, there are already attempts to eliminate genes of domestic horses from the remaining Przewalski's, and to breed out the gene that causes hernias of the diaphragm in golden lion tamarins. In more advanced mode, we may one day synthesise particular genes that we know are desirable but have been lost, and re-insert them into the breeding lines by genetic engineering. But the recreation of creatures *de novo* remains a nightmare vision. Our proper task must be to save the creatures we have now.

So we should ask, more soberly, what it is that technology may reasonably achieve in the next few decades – which (in the absence of effective technology) may be all the time that is left to many species. In general, these are the techniques that will lead us towards what has been called 'The Frozen Zoo'; in which gametes (eggs and sperm) and embryos are stored (probably in liquid nitrogen at $-196°C$) until such time as the curator wishes to quicken them into life, which he will probably do by placing them in the obliging womb of some surrogate mother. The Frozen Zoo, too, could conceivably be augmented by cloning which, as we will see later, does have some theoretical advantages, as well as obvious shortcomings.

The position in the early 1990s is that some of the contributory techniques are well advanced in some contexts and in some species, but none is universally (or even widely) applicable, and some of the necessary or at least desirable techniques are hardly even on the drawing board. But what, first of all, are the possible areas for intervention?

THE SCOPE FOR INTERVENTION

Most animals reproduce sexually; only a few have abandoned sex altogether. In all cases, sex involves the congress of two gametes: sperm from the male, egg from the female. They fuse to form a fertilised egg, otherwise known as a zygote, which is the beginning of the embryo.

After that, development follows many different courses, depending on the kind of animal. But in mammals (and most of this book is about mammals) the embryo first spends some time in the uterus of the mother, where it is nourished and increased in size many thousands of times. In marsupials the embryo is attached to the wall of the uterus only briefly: in eutherian mammals (rodents, ungulates, primates, etc.) it is implanted in the uterus wall. Birth takes place at various stages: some babies are very well advanced when they are born (such as herd antelopes, which are up and ready to run within a few minutes); while others are still effectively foetuses, (though once an animal is born it should not be called a foetus by definition). Marsupial babies in general are born at an extremely early stage of development (which essentially is still foetal).

After their birth, mammalian babies suck their mother's milk until they are weaned, which may take anything from a few days (some seals) to many months (kangaroos, humans). The suckling stage is essential both in eutherians and marsupials (and many a baby eutherian has died because the milk supply is inadequate). But whereas eutherians place roughly equal emphasis on the development in the womb and the development through suckling, in marsupials the balance is obviously tipped towards suckling. We should not underestimate the uterine phase in marsupials, however. The four-week period of uterine attachment in the kangaroo takes the embryo from a tiny speck to a baby mouse-sized creature able to crawl from the birth canal to the pouch; a prodigious journey, through the undergrowth of fur.

Clearly, there is a lot that can go wrong along the way. Humans are only one mammalian species among 4,000; but obstetrics, gynaecology, and neonatal paediatrics are among its most flourishing disciplines; what in the vulgar parlance of the 1980s would have been called 'major industries'. There are plenty of points, too, at which healthy and normal processes can be enhanced. In fact, there is an infinity of possible interventions, to correct malfunction at any point along the way, or to enhance processes that are already healthy and natural, but do not serve the present needs of other mammalian populations that are dwindling rapidly towards extinction. Thus, to discuss all that might be done would fill a university library. What follows, then, is a list of highlights from the spectrum of possibilities.

Fertility enhanced

A surprising number of males turn out to be infertile, or less fertile than they might be (subfertile). The problem is undoubtedly enhanced in populations that have been through genetic bottlenecks, and are inbred. Many cheetahs are infertile. Male infertility is a common problem in captive gorillas. Unfortunately, there is no clear correlation between fertility and dominance. Some macho males, keeping all other suitors at bay, have proved infertile.

Technologies are looming that might circumvent this problem, at least in the short term. For example, the deficiency often lies not in the nucleus of the sperm (which may indeed contain a perfectly good set of genes); but in the body of the sperm, the tail and the head, that are supposed to deliver those genes to the egg. Dr Harry Moore, at the Institute of Zoology of the Zoological Society of London,[1] suggests that in the near future nuclei from otherwise inadequate sperm might simply be injected into eggs.

Dr Bill Holt and his colleagues, who are also at the Institute of Zoology, have devised methods of diagnosing malfunction in sperm. They have long since found that appearances in sperm can be highly deceptive: that is, sperm with a high proportion of apparently malformed spermatozoa may later prove their ability to induce pregnancy, while good-looking sperm may somtimes fail. What counts more is motility; and in this (as recent studies on bull sperm suggest) stamina counts more than mere speed, for sperm that is still moving well after two hours is likely to be sound.

The problem however, is to measure the random movements of sperm on a microscope slide objectively. Subjective impressions can be highly deceptive – especially if the temperature is uncontrolled, for in the short term warm sperm moves more quickly than cold. The Institute of Zoology is now collaborating with transport engineers at Sheffield University, who have devised a computer that can measure and record the movements of individual vehicles in traffic. The problem of milling sperm seems much the same. Sperm motility should soon be measurable unequivocally. Of course, such diagnosis is not the same as treatment. But it is a first step. And it should prevent expensive disappointments.

In females, the ovaries operate in cycles. In mammals, one or several follicles first grow and mature within the ovary, over periods that vary from about three days (in ruminants and black rhinoceros) – each follicle being an egg surrounded by auxiliary cells (mainly granulosa cells) that secrete hormones. Then one or several eggs is/are released from the ovary – ovulation – leaving the granulosa cells behind; and these find their way into the reproductive tract where they may or may not be fertilised. The remaining granulosa cells then form a gland known as the *corpus luteum* (CL) which, *inter alia*, secretes the hormone progesterone, whose prime task is to prepare the uterus to accommodate the embryo. If the egg is not fertilised, and no embryo ensues, the CL regresses after a variable period (15 days in red deer and humans; 19 or so in black rhinoceros).

So – female fertility can be enhanced in a general way by treating with one or several of a variety of hormones (that may well include luteinising hormone (LH) from the pituitary gland – see below p.177) to cause them to mature and release more eggs than usual. This is *superovulation*. If the births are to be natural, then superovulation is a risky business, because animals in general are not geared to nourish and give birth to many more offspring than is characteristic of their species. In human obstetric circles, for example, potential triplets ring alarm bells. Superovulation is therefore generally combined with embryo transfer and/or storage (discussed below p.182) and (especially in the case of humans) may also be combined with *in vitro* fertilisation or IVF (also discussed, below p.180). 'Egg rescue' (again discussed below, p.179) vastly increases the number of eggs obtainable but is conceptually different and technically much more advanced than superovulation.

The next conceptual step in the general enhancement of reproduction is artificial insemination.

Artificial insemination

In artificial insemination (AI) sperm is taken from the male and inserted into the reproductive tract of the female. The sperm may sometimes be used fresh, in which case it generally has to be used within a few hours – although pig sperm will stay fresh and viable for several days, and commercial pig breeders commonly send boar sperm through the post. Or (which in general is much more convenient) the sperm may first be stored, generally in liquid nitrogen at $-196°$ C, and then thawed before use. Timing is important. The female has to be about to ovulate, or to have just ovulated: the optimum time of insemination relative to ovulation varies from species to species.

The advantages of AI are obvious. Sperm is much easier to transport than entire animals; particularly entire mature male animals. Animals are traumatised if they are moved from place to place (and so, sometimes, is the herd they leave behind). Sperm once frozen costs much less than an entire animal to maintain; and can be kept for much longer. It seems likely that deep-frozen sperm might be kept indefinitely; perhaps for centuries. In principle, sperm (and therefore the genes) of every captive male mammal that is now alive could be kept in deep freeze in perpetuity, within just a few rooms. And why stop at captive animals? It is now feasible (and in some cases straightforward) to anaesthetise animals in the wild without doing them permanent harm; and while they are anaesthetised, sperm can be collected from them. Thus we could in principle rescue the genes from individuals from every population of endangered large mammal that is still extant.

There are still many problems, however. In many species, for example, it is difficult to collect sperm. In principle, it is best to do this while the animal is conscious. This is obviously applicable primarily to domestic animals, and even here, it is not easy. Domestic cattle are the species in which AI is most advanced; and in them, the bull may be induced to 'serve' a dummy cow, ejaculating into a giant condom. Red deer are catching on as a domestic species, and AI in them is highly desirable to effect rapid genetic improvement, because stags left to themselves will 'cover' only about 25 hinds per breeding season (which means per year). The technique now favoured in New Zealand (where red deer farming is most advanced) is to fit a condom into the vagina of a hind, though it is difficult to prevent the condom from turning inside out as the stag withdraws. As stags produce only about one ml of semen per ejaculate (at most), it is easy to lose the entire load. An alternative is to wait until the stag is about to penetrate the hind, and then slap the sheath over his penis by hand. Red deer stags in rut are formidable beasts, however. This is not an exercise to be undertaken lightly.

In practice, semen from non-domestic species, whether captive or wild, is best taken while the animal is anaesthetised. Ejaculation is then induced electrically (generally via the rectum) or by hand. Anaesthesia itself (with anaesthetic dart) is relatively straightforward these days, but there are many problems none the less. Rhinos, for example, in general refuse to ejaculate when anaethetised, or if they do, they produce semen of very poor quality (containing seminiferous fluid but very few sperm). Perhaps there is a fault with the anaesthetic; again, scientists at London's Institute of Zoology are working on ways of anaesthetising rhinos (and other species) without turning them off completely.

In all species, AI may fail first time around. Domestic cows are content to endure repeat inseminations; and women in fertility clinics are highly motivated volunteers, also willing to undergo multiple inseminations over several days. Insemination in exotic animals generally has to be done under restraint (which would generally involve various degrees of sedation, albeit falling short of surgical anaesthesia). Repeat performances are highly undesirable, even if the semen was available.

Neither is it necessarily easy to freeze sperm – which for many purposes is highly desirable, even if it is not always strictly necessary. Sperm in general do not take kindly to cooling. Granted, male mammals commonly keep their testes cool (at least in the breeding season) by holding them away from the body cavity in a scrotum. But the temperature of a sheep's scrotum (say) is only around 32° C; a few degrees below body temperature. Sheep sperm is damaged if cooled to 5° C. But 5° C is nothing compared with −196° C. Sperm tolerates such prodigious freezing only if it is protected within some suitable medium, and only if freezing – and subsequent thawing! – are carried out at an appropriate rate.

However, sperm from different species vary enormously in their tolerance of freezing; and the kind of media that will (or in many cases will not) protect them when they are cooled also vary enormously. Thus, in general, cattle sperm are reasonably robust, but sheep much less so. Antelope, evolutionarily speaking, are somewhere between cattle and sheep: some are cattle-like, and some are more ovine. Correspondingly, some antelope sperm freezes more easily than others, though Bill Holt and his colleagues are confident that they should now be able to freeze most antelope sperm successfully. Pig sperm so far freezes badly, however. It is lucky that it lasts so well when fresh (though this is of limited value for conservators of exotic pig species).

Chimpanzee sperm freezes well (relatively speaking). Of course, chimps are very closely related to humans, whose sperm also freezes well; but this may not be the point, for there can be significant differences even between closely related species. Probably the point is rather that chimps have received a great deal of attention, because of their role (regrettably) as laboratory animals. Success with gorillas (also very closely related to chimps and humans) is much more equivocal. Monkey sperm has not frozen well so far. Among carnivores, dog sperm clearly freezes more easily than cat. Giant panda sperm is robust (and a cub has been born in Milan using frozen sperm), but red panda sperm much less so. Among birds, sperm has been frozen successfully from domestic poultry and some birds of prey, and there is much interest in freezing sperm of endangered parrots. Budgerigar sperm has been frozen successfully at the Institute of Zoology.

This may seem like a pretty arbitrary list, not to say random; and so it more or less is. There are areas of biology that are excellent science, and others that are simply a series of recipes. This latter has been the general state of AI technology. Because the whole issue is so complex; because there is very little hard science to guide our hands, it has taken more than 40 years to devise recipes and techniques to achieve even the limited (though admittedly adequate) success we perceive in domestic cattle. To devise suitable methods for all 2,000 endangered species of land vertebrate at the same rate would take – well, more time than there is.

Males, however, are only one half of successful AI. The other half is the female.

The distaff side of AI

Some animals do not mate until just before ovulation; some more or less at ovulation; some after. There are also some variations on this theme. Some ('induced ovulators') are prompted to ovulate by copulation; they include squirrels, beavers, racoons, dogs, weasels, and many others. Female bats keep the sperm in store (for up to 198 days in some species), and ovulate some time after mating. The females of the short-tailed opossum, *Monodelphus domestica*, a South American marsupial, do not even begin the follicular phase of the cycle until a male comes among them. Until then, they are sexually quiescent.

The only point here, however, is that each species presumably knows its own business best; but in general, among mammals, sperm do not remain viable in the reproductive tract for more than a day or so (bats aside), while eggs, too, rapidly lose their 'willingness' to be fertilised. Practitioners of artificial insemination must therefore know when the female is ovulating: or, better still, predict the time of ovulation.

The females of some animals come clearly 'into heat' at the time of ovulation, so that even a human can see that they are receptive. This is generally true of domestic cats and dogs. Many other species are far more reticent. In many the signals are olfactory; the females have a particular scent at the time of ovulation, which the males detect, often by sniffing the urine. In some the effect is pheromonal: that is, the chemical signal is not merely a scent, but acts directly as a hormone to influence the behaviour of the male. Farmers wishing to carry out AI often detect ovulation in their cows or ewes with the aid of 'teaser' bulls or rams – who are able to tell when females of their own species are ovulating, and reveal this to the farmer by their own behaviour.

To the breeder of exotic animals, however, such options are often not available; a teaser bull elephant, for example, is a bizarre notion. Indeed, breeders have four main recourses. The first is to trust to luck; to take a reasonable guess at when insemination might work, and go for it. As we have seen before in this book, a lot may be achieved by common sense and a general sense of good husbandry. Much better though, is the second option: to know the physiology of the animal; to know how long the cycle is; when, in the cycle, ovulation occurs; and when, ideally, relative to ovulation, the sperm should be introduced. Such knowledge is of limited use, however, if it is merely theoretical. The third requirement, then, is to be able to monitor the ovarian cycle, and detect the time of ovulation; and, better still, to be able to predict the time of the next ovulation from a knowledge of the time of the previous ovulation, or the time of the end of the previous cycle.

Finally, in an ideal world the breeder would be able to control the cycling of the female, so that he can ensure that she ovulates when the sperm is good and ready. If he is using the sperm fresh (in the absence of reliable freezing technology); and if he is importing the sperm from some other zoo (as will very possibly be the case), then close coordination is mandatory. Ideally, too, all the monitoring and control of cycling should be exercised with minimum disturbance to the animal.

Progress in these various directions can again be illustrated by work at London's Institute of Zoology. At the Institute, for example, Cheryl Niemuller has been trying first to 'characterise' (that is, to describe in detail) the ovarian cycles of elephants; and second to find practical ways of monitoring those cycles. In general, ovarian cycles can most easily be monitored by measuring the rise and fall of hormones in the blood. In particular, the follicle is prompted to grow by pulses of follicle stimulating hormone (FSH) and luteinising hormone (LH) from the pituitary gland at the base of the brain. As the follicle is growing, the granulosa cells produce oestrogens in increasing amounts, and small amounts of progesterone. Just before ovulation (in general) there is a surge in output of luteinising hormone (LH) from the pituitary gland beneath the brain, which stimulates ovulation. After ovulation, when the follicular cells have become the cells of the CL, progesterone output increases markedly, and oestrogens decline. Thus (in general), we can tell when the follicular phase comes to an end by a peaking and then a fall of oestrogens; when ovulation is about to occur by a surge in LH; when it has just occurred, by the beginning of the rise in progesterone; and when the second half (the luteal phase) of the cycle has come to an end, by a falling-off of progesterone. Various hormonal changes signal the beginning of pregnancy, including (in general) persistence of progesterone. So – all the curator needs to do is to follow the rise and fall of hormones in the blood.

There are snags, however. One outstanding drawback is that few exotic animals will stand still and allow blood samples to be taken. Among large exotic mammals, elephants are almost unique in this respect. They have a history of being handled, and some individuals, at least, are tractable. Thus Cheryl Niemuller has been able to take serial blood samples from elephants, and is well on the way to describing the ovarian cycle. However, the changes in hormones do not necessarily denote the time of ovulation exactly; or at least, the time of ovulation cannot be inferred simply from the hormone changes, unless you know the physiological details of the animal in advance. But the time of ovulation can be directly perceived by ultrasound; which can show the characteristic changes in the surface of the ovary as ovulation occurs. Generally, however, the ultrasound probe is inserted through the rectum, and the ovaries are scanned through the wall of the gut. But an elephant rectum is extremely muscular; the gut wall is thick; and the distance from the anus to the ovaries is greater than the length of a vet's arm. Thus, it is very difficult to get ultrasound pictures of an ovulating elephant ovary. It is not yet possible, therefore, precisely to relate the changes in elephant blood hormones, to actual events.

In addition, although some zoos (such as England's Chester) are equipped to take at least weekly blood samples, most curators would, in general, prefer not to try. Besides, weekly samples are probably not quite frequent enough to detect ovulation (though they will show whether or not the ovaries are functional, and hence whether the animal is fertile). Neither would it be possible to take blood samples in the field, and thus find out whether particular wild elephants were fertile, or were pregnant. Hence, it is desirable to be able to detect hormone changes not in the blood, but in the urine; and preferably also in the faeces (which in general are easier to collect than urine).

This raises new problems. Assays for detecting hormones in media that are as dilute and biochemically complex as urine have to be extremely sensitive and specific; and you cannot make a specific, sensitive assay unless you know exactly what you ought to be looking for. A difficulty here is that the hormones that control ovarian cycling may differ somewhat in chemistry from species to species. In addition, hormones do not in general appear in the urine unchanged; usually (though not invariably) they are first broken down by enzymes in the liver, so that what finally is excreted is not the hormone itself, but a hormone metabolite. Different species – even closely related species – may employ different liver enzymes for the breakdown, however, so that even if they employ the same hormones in their ovarian cycles, they may produce different urinary metabolites.

Thus in elephant urine, Cheryl Niemuller has found that the principal breakdown product of progesterone is pregnanetriol. In women, progesterone is broken down to pregnanediol. Indeed, women produce pregnanetriol only if they are suffering from a disorder of the adrenal cortex, known as congenital adrenal hyperplasia: a condition that results in virilism. There is no standard assay for pregnanetriol (since it does not feature greatly in human medicine) so one of Cheryl Niemuller's tasks was to create one.

It is highly desirable, as we have noted, to control the cycling of females to some extent; to bring them into oestrus at a specified time. One general method for doing this is to bring the previous cycle to an end by administering prostaglandin hormones, which cause the CL from the previous cycle to break down, and allow the next cycle to begin. Provided the length of the follicular phase is known (and we have seen that it is highly variable, even between ostensibly similar species) the time of ovulation is predictable. The point is made, however. AI is as much a matter for females as it is for males.

The next logical step in the enhancement of fertility is to increase the total yield of eggs from the ovary – not simply to superovulate, but to 'rescue' eggs that would not normally even be allowed to mature.

Egg rescue

At birth – or even before – the ovaries of mammals contain all the eggs they are ever going to contain. No more appear during life outside the womb. In fact there is a steady loss. Even before the female matures and begins to release eggs in the course of normal ovarian cycling, some of them begin to decline (by a process called atresia), and this apparently gratuitous waste continues throughout life. When the female begins to cycle, the loss through ovulation adds to the atresia. Finally, at the time that in humans is called menopause, there are no more eggs left. Most wild animals in practice die before they reach this 'post-mature' state; but a few, like some lucky elephants, stay around, and remain important to the herd as elder stateswomen, aunts, and protectors. (This is not fancy. We saw the value of older animals who perpetuate 'traditions' in chapter 5).

At the time of her birth, the ovaries of a mammal may indeed contain contain 1 million eggs. An animal like a rhinoceros – like a human – is unlikely to give birth to more than half a dozen calves in her life, in the wild; and very few captive rhinos have produced as many. The waste, therefore, is prodigious.

But surely the waste could be prevented? Surely if a female dies or is killed or culled before she is post-mature – even indeed if she is still an infant, and perhaps dies at birth – her ovaries could be collected, and the eggs taken out and rescued? Javan rhinos are down, now, to about 50 individuals. Yet the remaining females between them might contain a million eggs (and that would be a conservative figure). And one male could in theory supply several billion sperm, so that should be no problem (once the practical problems have been overcome). A million Javan rhinos would probably be more than have ever existed at any one time in the past. In practice, 'egg rescue' is beginning to happen. The pioneers are at IAPGR in Cambridge, where Dr Bob Moore has so far 'rescued' the eggs of mice, and recently of sheep. We can assume that one day it will indeed be possible to rescue the eggs of any species, effectively at will; including the Javan rhino.

But for the moment there are enormous snags. The eggs present in the young female – and most of those present in the ovaries even of older females – are 'immature'. Before they can be fertilised, their chromosomes must divide (by meiosis) to produce true gametes, with only half the number of chromosomes of the parent, this being a two-stage process. In addition, the cytoplasm must undergo many maturational changes. The various stages of development are acknowledged by terms such as 'primary oocyte' and 'secondary oocyte'.

So far, the precise nature of the maturation process is not understood. Neither, therefore, do scientists know the precise conditions that prompt those changes to take place. So far, then, the process of egg maturation *in vitro* is at the same general stage, though not so advanced, as the freezing of sperm: there is a series of recipes that succeeds, in some instances, but nobody properly knows why.

At the Institute of Zoology, Dr Helen Shaw and her colleagues (as well as Bob Moore in Cambridge) are seeking to put the whole business on a more scientific basis. It is known, for example, that in addition to nutrients, eggs need to be exposed to various hormones and growth factors – which again will differ from species to species. In practice, various hormonal stimuli evidently work upon the follicle cells, which in turn influence the maturation of the eggs.

Thus, maturation clearly depends in part upon secretions from the granulosa cells that surround the egg in the follicle. These secretions include steroids, such as oestradiol. Helen Shaw has found that secretion of oestradiol in the maturing follicle is in turn stimulated by insulin-like growth factor (IGF1), which is a small protein. IGF1 given to mature granulosa cells immediately promotes steroid secretion; but immature granulosa cells do not respond to IGF1 until four to six days have passed and must first be treated with oestradiol.

Clearly, the factors that will do the trick will vary from species to species; but once the general range of patterns is understood then, as with sperm freezing, it should be possible to home in on the optimum conditions routinely. In practice, biologists are unlikely to be able to handle more than a few hundred eggs from any one female: selecting the oocytes that are closest to maturation. But a few hundred per female would be enough.

Eggs could in principle be frozen. But the more probable course is to fertilise them *in vitro* as soon as they are mature and then to freeze the resulting embryos. So: in vitro fertilisation is the next logical step in this discussion.

In vitro fertilisation

The first to show that human babies could be conceived in a 'test tube' (or, more usually, in a petri dish) were Dr (now Professor) Robert Edwards from Cambridge University, and the late Dr Patrick Steptoe, from Oldham in the north of England. Edwards was the scientist; Steptoe, an obstetrician, was the clinician. The first 'test tube baby' was born in 1978. By the end of the 1980s, there were more than 10,000.

In general, IVF is again in much the same position as sperm freezing. Scientists know that it can be done; they have the required techniques for a few species – which of course include human and some domestic species, with a few exotics so far achieved here and there. Egg rescue and *in vitro* maturation of eggs are not necessary precursors of IVF (which is just as well, because it is more difficult than IVF itself) – though eggs fertilised *in vitro* may in practice undergo the final stage of maturation *in vitro*.

Normally, however, the ovary is first induced to superovulate by one of various hormone treatments. Then the obstetrician (or vet) employs the technique of laparoscopy, the development of which was one of Dr Steptoe's important contributions. He makes a small incision in the abdominal wall; distends the abdomen slightly to give more room, with an innocuous gas such as carbon dioxide; and inserts a multi-purpose probe which, *inter alia*, carries light down a bundle of glass fibres ('fibre optics') and thus illuminates the scene. When follicles are ready to release their eggs, they are swollen, and stand out clearly at the surface of the ovary, like tumuli. The contents (including the egg) can then be aspirated (sucked out). Such more-or-less mature eggs are employed for IVF.

Sperm, though, is not ready to fertilise an egg when it first leaves the testis. For the final stage of its 'maturation' it needs to be 'capacitated'; which involves changes in the outer membranes that allow the final penetration of the egg. Capacitation generally takes place in the female tract: so conditions that allow it must be simulated *in vitro*. Such technical problems are not great, however.

Probably the greatest single problem for IVF in exotics are the essentially veterinary ones. Eggs can be obtained for laparoscopy only under general anaesthesia. All anaesthesia carries some risk, and individuals of endangered species in particular should not be risked. It may also be difficult (with present technology) to supply sperm as required, as we have already seen. Vets perceive ethical problems, too; for in general such treatments should be carried out only on volunteers (as with humans) or for the benefit of the patient. Is a (non-volunteer) healthy female a patient? Does she herself benefit from the IVF? Taken all in all, though, IVF must be seen as an important technique on the path to the Frozen Zoo; and hence, it is an important component of conservation technology.

In humans, embryos would be produced by IVF mainly to overcome some difficulty of natural fertilisation, or as a means of diagnosing suspected chromosome or genetic disorder before pregnancy begins; and the embryo so produced would be transferred into the womb of its own mother. So far the success rate is less than desirable; probably only about 15 per cent of transferred pregnancies lead to pregnancy, although Dr Edwards feels that 40 per cent ought to be achievable. However, only about 25 per cent of *natural* matings performed at the time of peak fertility in humans lead to pregnancy (and about half of all human pregnancies abort naturally before term). Humans are not a very fertile species. It's just as well that we live a long time, and take good care of our offspring. IVF in animals might, in general, produce a higher rate of success. In many species, after all, the conception rate at peak fertility may be well over 80 per cent.

But this brings us to the next major set of technologies: embryo transfer (to the womb of the natural mother, or to a foster mother); and embryo storage.

Embryo transfer; embryo storage

Inevitably (the point seems to go without saying, in the field of applied reproductive physiology) techniques for embryo transfer are most advanced in humans and cattle. In humans, embryo transfer almost always follows IVF; IVF that has been carried out to overcome some problem with natural conception, or to provide a means of diagnosing some anticipated genetic or chromosomal disorder. After IVF, two or three healthy embryos are generally transferred to the womb of the natural mother. Surrogacy in human motherhood is still highly controversial.

In cattle, the point of embryo transfer is not to overcome fertility problems, for no farmer would want to breed from a cow (or a bull) that was not extremely fertile. The point, rather, is to obtain many more embryos from an elite cow than she is capable of carrying to term herself. The embryos are not produced by IVF. Instead, the mother is superovulated; the eggs are fertilised in utero by AI (conception does not have to be by AI, but this is normally the case); and the young embryos are washed out of the reproductive tract and each one is implanted into a foster, or surrogate mother.

The use of embryo transfer to increase the reproduction of favoured cows undoubtedly works. Cows have a nine-month pregnancy, and – at best – naturally produce only one calf per year. But if they are not made pregnant, they cycle again a month later. In principle, if cows are superovulated and covered by AI, a half dozen or so embryos can be taken from them every seven weeks or so (though 90 days is more likely). In practice, then, cows commonly produce 12 to 24 calves per year by this method, and have been known to produce up to 50.

Embryo transfer in exotic animals is much more similar, in practice, to cattle than to humans. That is, there is in general no need to employ IVF (though IVF might be involved under special circumstances – and would of course be the norm if and when egg rescue comes on line). Normally, young embryos are simply washed out of the wombs of their natural mothers, and then transferred into foster mothers. Again, the general idea is to raise the output from desirable mothers.

However, there is, of course, an enormous contrast in philosophy between embryo transfer for the conservation breeding of exotic animals, and embryo transfer in domestic cattle. The general aim of conservation is not to produce vast numbers of 'elite' animals, but to ensure that the maximum range of genes is conserved. None the less, as we saw in chapter 4, when an animal is extremely rare, the niceties of breeding strategy have to give way, to some extent and in the short term, to the simple need to keep the population going. There is some safety in numbers, even if the offspring of only one individual are 'over-represented'. Besides, as we also saw in chapter 4, there is inevitably a loss of genes with each generation, because each gamete contains only half its producer's genome; but the loss is less, if the animal has a lot of offspring. So if a species was down to its last few individuals, there would be a very strong case indeed for boosting output as much as possible – provided, of course, there were suitable foster mothers. No one would complain (would they?) if any of the few remaining Javan rhinos could be persuaded to produce 20 or so calves, even if they were all a little genetically uniform. In addition, females that were not themselves genetically ideal (being too similar to others) could still be valuable as foster parents in their turn: bearing the young of more (genetically) valuable individuals.

In addition, we could envisage that genetically valuable individuals of rare species might be helped through any infertility problems in the same way as humans are. Infertility, after all, is not generally hereditary; it is much more likely to be caused by some infection, for example. Even the relative infertility that can be brought about by inbreeding could in principle be cleared up in a generation; that is, the heterozygous, outbred offspring of two unrelated subfertile homozygous parents should not itself be infertile.

Furthermore – and most intriguingly – embryos of exotic animals should often, most conveniently, be transferred into the foster mothers *of different species*. After all, some species that are now extremely rare have close relatives that are extremely common; notably the world's wild cattle, several of which are extremely rare. Friesians (in the United States generally known as Holsteins) can be (and in a few cases have already been) excellent foster mothers.

Historically, embryo transfer was developed long before IVF. If it had not been so, then there would have been little point in developing IVF, for the resulting embryos would have had nowhere to go. Indeed, the first successful embryo transfer was carried out in Cambridge, England, in the late 19th century, in rabbits. Then came attempts with sheep and goats in the 1930s, in the United States; then cattle, at Cambridge again, in the 1940s; and – finally, when all the subsidiary techniques were more or less in place, including the synchronisation of ovarian cycles – general methods for use in cattle were developed at Cambridge in the 1960s. A variety of techniques began to be applied increasingly routinely in the 1970s; and by 1985 at least 100,000 calves had been born following embryo transfer. Most have been carried out in highly commercial herds in the US and Europe, but here is a (relatively) high technology that could be of particular value in the Third World, where millions of tough but unimproved cattle could do with rapid upgrading.

In cattle, the whole sequence of embryo transfer typically goes as follows, though many variations are possible. First, the oestrus cycles of the donor and the recipient are synchronised by giving hormones (for example prostaglandins, to bring the previous cycle to an end, and ensure that they both begin the next one together). It is important that the uterus of the recipient is at the precise stage appropriate to receive the foster embryo at the time it is due for implantation (which is when it is a blastocyst – a hollow ball of cells, still the same size as the original fertilised egg).

A few days after AI (timed to coincide with ovulation), the young embryos are washed from the ovarian tubes or the uterus of the donor. This can be done surgically: for example, passing a cannula to the top of the ovarian tubes; or it can be done non-surgically – squirting culture medium into the uterus through the cervix, and then floating out the embryos, just before implantation.

Embryo freezing, yet again, is at the same kind of state as that of sperm freezing. It can be done (and in general embryos seem more robust than gametes); we can reasonably expect that any kind of mammalian embryo could be successfully frozen, given a few months research on each; but the technology cannot yet be classed as an exact science. As we will see, however, there is now a growing cadre of young animals, of several species (including cats and antelopes) that were deep frozen before they were implanted. This particular technology should not delay the Frozen Zoo.

From all that has been said so far, the general problems in applying these methods to exotic animals are obvious. However gently the procedures are carried out, exotic animals in general require anaesthesia, which is potentially dangerous. In most cases, the reproductive cycles are not well understood; so timing is difficult. None the less, considerable successes have already been achieved; not many numerically, but enough to offer hope that embryo transfer has enormous and wide potential.

Outstanding in this field is the Center for Reproduction of Endangered Wildlife (CREW), founded at Cincinnati Zoo (as a collaborative venture with the University of Cincinnati and the nearby King's Island Wild Animal Habitat) in 1981. Their successes, under the zoo's director Dr Ed

Maruska and CREW's Research Director Dr Betsy Dresser, reflect (and in many instances lead) world progress so far.[2] Thus, specifically in the field of embryo transfer, CREW achieved the world's first birth of an exotic animal in 1983. This was carried out by non-surgical transfer in eland antelope: a common species, but potentially able to foster other large antelope, several of which are severely endangered.

One year later – 1984 – CREW achieved the first transcontinental embryo transfer with exotic species: the embryos were brought from Los Angeles to Cincinnati. These were highly significant events for other reasons. The embryos were bongos: rare, beautiful, lowland forest antelopes from across the middle of Africa. One young bongo was successfully born to a bongo foster mother; the other was fostered by an eland. In 1984, too, Cincinnati achieved the first successful transfer of a frozen embryo in an exotic species: again, an eland. The first cats were born from transferred frozen embryos in 1986. These were domestics; but of course, many of the world's 36 or so species of cat are endangered, some severely so; particularly the small spotted cats, hunted for their fur.

1987 saw the non-surgical embryo transfer of a gaur embryo into a domestic Holstein cow. Gaurs are truly wonderful animals from Asia, classified as 'vulnerable', and so big (the males up to two metres high at the shoulder) that they make you gasp. The following year – 1988 – Cincinnati produced an eland calf from only *half* an embryo. The other half is still frozen. Thus we are at the beginning of cloning, to be discussed below. 1989 was another *annus mirabilis*: an Indian desert cat, born to a domestic cat. This was the first exotic cat produced by IVF; and the first transfer of a cat embryo between species.

Betsy Dresser's chief ambition at present – the beginning of 1991 – is to employ these techniques among the world's diminishing populations of rhinoceroses; perhaps, for example, to transfer embryos of the highly endangered black rhino, and the even more endangered Sumatran, into white rhinos. Non-surgical techniques in these animals seems mandatory. There is a long way to go, but within the US there are a lot of resources, with nearby King's Island, and a few Texas ranches, able to maintain good-sized herds; and zoos such as San Diego (in particular) providing considerable expertise. The kind of physiological information now coming out of London is also of obvious importance. Rhinos in general are extremely parlously placed; and interventionist technologies may well play a part in their salvation, if they can be saved at all.

The more we know about the mechanisms of implantation, the more successful embryo transfer is liable to be; and the greater the possibility for transferring embryos even into species to which they are not particularly closely related. Thus, elands have already served as fosters for bongos. Why not sables? Or scimitar-horned oryx? Recently, too,

American scientists attempted to transfer 24 embryos of thinhorn sheep into domestic sheep. (Thinhorns are *Ovis dalli*: wild sheep from North America otherwise known as Dall's sheep). None of the transferred embryos survived to term. With more knowledge of the mechanisms of implantation, the uteruses of the mothers might have been made more receptive. The transfers might have succeeded. How widely might the net be spread, if only we understood at the molecular level, what implantation involves?

There is one further, and very exciting possible refinement. At Cambridge, Dr Chris Polge has been developing methods for inducing animals to accept embryos from species that are not very clearly related, by making chimaeras. This we should discuss.

Chimaeras

Chimaeras were pioneered by Chris Polge and his colleagues, when he worked at what was then the Animal Research Station at Cambridge. The basic idea is simply to mix cells from two different, very young embryos (blastocysts); and the resulting mixed embryos can then be re-implanted into a complaisant uterus, and grow into a mixed animal. Remarkably, this can produce viable embryos that grow into mature animals even if the two embryos are of different species; provided (one assumes!) that the species are not *too* distantly related.

Thus, when the cover of *Nature* carried a picture of a sheep-goat chimaera,[3] some newspapers delighted in giving it a name: a 'geep', or a 'shoat'. This would have been appropriate, however, only if the animal had been a true hybrid; that is, if *each* of its cells had carried a set of chromosomes from a sheep, and a set from a goat. Chimaeras, however, emphatically are *not* hybrids. Part of this pioneer animal was pure goat; and part was pure sheep. Overall, the chimaera looked like a somewhat lanky and rather scruffy sheep, which could not quite make up its mind whether it wanted to be woolly or not.

Such a production may seem to be merely a piece of indulgence. Not so. For one thing, chimaeras offer a way of investigating the vexed question that we have already outlined. How does the uterus 'decide' that one kind of embryo is acceptable for implantation, and another is not? What signals pass between the embryo and the uterus? With answers to these questions, we should be able to enhance implantation: reduce the failures that may follow embryo transfer; widen the range of embryos that any one foster mother may accommodate. They also offer ways of studying development: how and at what stage particular cells in an embryo 'decide' that they will become part of the nervous system or a muscle; and how they then differentiate and move to the appropriate positions. I will look at this issue again later. In addition, however, chimaeras could at least in principle be applied directly to conservation breeding.

The first and obvious point is that foster mothers are likely to be able to raise chimaeras between embryos of their own species and those of other species, when they would not raise a pure embryo of the other species. Thus embryo transfers of pure goat into pure sheep, or vice versa, do not work.

The acceptibility of the chimaeric embryo can be further enhanced: in theory and (Dr Polge has already shown) in practice. Thus, as a mammalian blastocyst develops, cells divide at an early stage into two lines: one that begins as the inner cell mass, which will become the embryo proper; and one that becomes the cells of the trophectoderm, which in their most mature stage provide the placenta. Only the cells of the trophectoderm are in contact with the mother's uterus. If the mixing is carried out at precisely the right time, it is possible to create a chimaera that contains only inner-cell-mass cells from one species, and only trophectoderm from the other. Provided the trophectoderm cells are of the same species as the mother, the mother should accept the embryo 'unquestioningly'. The inner-cell-mass could be of a species that she would not normally accept: but when it is born, it would be *pure*.

The technique could produce pure offspring from otherwise in-hospitable mothers even without such elaborations. As we have seen, parts of a chimaera contain pure cells from one parent; and parts contain pure cells from the other. Thus, in principle, sheep-goat chimaera could contain gonads that were pure sheep or goat; and could therefore produce pure gametes. Thus, a rather messy-looking chimaera between a rare antelope and a common antelope could be born to a common antelope; but that chimaera (with luck!) could produce gametes of the rare antelope.

To be sure, it is not easy to envisage that chimaeras in practice will play much part in the conservation of rare species. It will be some years (perhaps decades) before it would be reasonable to apply such techniques to the Javan rhino (say); as we have already seen, there are a great many hurdles to be crossed with rhinos before we get near to such refinements. But within a few decades, either the Javan rhino should have been made safe (by more conventional means), or it will be extinct.

But chimaeras are a shot in a locker; and as we have seen, the study of them could hasten the success of less spectacular technologies. Chimaeras, in fact, emerged from studies of a conceptually simpler technology, which has more direct conservational relevance: cloning.

Cloning

A clone is a group of organisms that is produced asexually from individual cells (or larger parts such as cuttings, in the case of plants), from a single parent. Thus, all the individuals in a clone are genetically identical to all the others (and to their parent). But there can be a great many of them; for example, all the Cox's Orange Pippin apple trees that have ever lived are a clone from one original tree that was first grown (from a pip; a conventional product of sexual reproduction) in the 19th century.

Plants are very good at asexual reproduction (a point I will again return to). So are bacteria, fungi, protozoa, and a few multi-celled animals such as corals and sponges. Many other animals (including bees and aphids) practice parthenogenesis, in which offspring arise from unfertilised eggs. This is asexual reproduction; but it does not necessarily result in true clones, because unfertilised eggs are gametes and gametes are not genetically identical one to another. Each gamete contains only a selection of its parent's genes, and no two gametes (broadly speaking) contain the same selection (unless the parent is extremely homozygous).

However, even animals can produce true clones naturally, albeit generally of extremely limited size. Thus in nine-banded armadilloes the four-cell embryo invariably divides to produce four separate cells, each of which grows into a new individual; thus nine-banded armadilloes invariably produce identical quads. Even humans occasionally produce identical twins by the same means; the two-cell embryo divides, and each half becomes a complete embryo. Note that since identical twins must be genetically identical, they must also be of the same sex (provided we are talking about mammals; in some other animals – such as tortoises and many fish – sex depends on upbringing). There are numerous examples in literature of identical twins of different sexes: Viola and Sebastian in Shakespeare's *Twelfth Night* (the plot doesn't work *unless* they are identical); Nick and Catherine in Iris Murdoch's *The Bell*. But then, you don't have to be a biologist to be a writer.

However, it is already possible to produce small clones of mammals by separating the cells of very young embryos that have just begun to divide. Thus at Cincinnati, Betsy Dresser produced an eland calf from only half an eland embryo. When I visited her in the autumn of 1990, the other half was still in store, frozen. By the early 1980s Chris Polge had already produced five identical lambs, each from an individual cell from an eight cell embryo.

In fact, by the simple division of an embryo, a clone of eight is as large as is liable to be obtainable from a mammal. The point is as follows. When an embryo consists only of one cell, its genes are switched off. The first cell divisions take place without their switching on. They are orchestrated by the cytoplasm of the egg; and the cytoplasm is, of course, donated entirely by the mother, and therefore is maternal tissue.

Not until after the first, second, or perhaps the third cell division – the exact stage varies with species – do the embryo's own genes switch on, and begin to produce their own proteins. This is obviously a very significant developmental step (one being studied by Dr Leeanda Wilton at London's Institute of Zoology). Unless the genes switch on, the embryo fails. A failure to switch on may be responsible for some, perhaps many, natural failures. It is one of the reasons why (so far) it has been difficult to culture the embryos of species such as pigs *in vitro*; and unless embryos can be cultured *in vitro* to the blastocyst stage (which involved several divisions, producing many hundreds of cells) they cannot be implanted into a uterus.

But animals play a price for the switching on. Until they are 'awakened', the genomes within each cell of the young embryo are totipotent: that is, given the right conditions, they are able to give rise to successive generations of cells which between them can differentiate into *all* the different tissues characteristic of the species. That is, the single cell can give rise to an entire viable organism. But once they are switched on, they begin to lose this totipotency. It happens to be the case that in ruminants (such as sheep), the switch-on does not occur until just before the fourth cell division (leading to a 16-cell embryo). Thus, with ruminants, it is theoretically possible to produce identical octuplets, by separating eight-cell embryos soon after the third division. Chris Polge's quintuplets come very close to this. But in species that normally switch on earlier than this, only smaller clones can be produced – at least by simple embryo division.

Again, we may note that a clone does not fulfil the conservation breeder's ideal – which is to produce populations of animals that between them contain the maximum possible genetic diversity. The members of a clone contain only one genome – at most, only two alleles of each locus – repeated many times. Yet, as we have suggested earlier, a clone would still be useful. If the present tiny population of Javan rhinos were repeated exactly, two or three times – with another absolutely identical herd in Europe, and another in the US – we could feel easier about its future, if only in the knowledge that the poaching of any one individual would not be as irrevocable as it would be now. If any one female were cloned half a dozen times, or a hundred times – each would be a vehicle to bear the offspring of a greater number of males. As we noted, too, in chapter 4: the number of genes lost per generation is less, if more offspring are born. The offspring of a hundred identical females would retain a greater proportion of the genes of their mothers, and their various fathers, than the five or six offspring that might be born to any one of them. Whatever the genetic quibbles, then, when an animal is in deep trouble, clones could be extremely useful.

This discussion takes us to the edge of foreseeable possibilities; probably (just to pluck a time-scale almost at random) to the edge of feasibility for the rest of the 20th century. But it is worth speculating briefly, as to the further future.

THE FUTURE

In the very long term it might be possible to restore totipotency to animal cells that have already differentiated. If this could be done; and, more, if individual genes could be switched on or switched off at will – then true cloning on a truly vast scale could be practised. I stress 'very long term' because at present, the idea of exercising the degree of control over animal genomes that is implied seems remote indeed. Some animal genes can be turned on or off at will; and indeed, many drugs (or hormones) operate effectively by turning on genes, or switching them off again. But the possibility of restoring (say) the ability of a muscle cell to act like a liver cell seems extremely remote. There seems no reason why it should not be made to work; it offends no 'bedrock laws of physics'. But biology at present just is not that accomplished.

Herein lies one of the most profound differences between animals and plants. You can take virtually any live cell from any part of a carrot (say) or a tobacco plant, and culture it to form an entire plant. Even after differentiation the cells reveal (with just a little coaxing) that they retain the quality of totipotency. Indeed, you do not need to be a botanist to achieve this. A gardener who takes a cutting, and roots it, reveals the potential of the cells he exposes. Not all plants are equally totipotent, however. For example, it is easy to produce huge clones from single cells of carrots (if you should want to do such a thing); but very difficult to do this by culturing coconut palms (though this would be highly desirable, as coconuts cannot be cultivated by cuttings, and are difficult to improve by conventional plant breeding because of the length of time between generations).

We may speculate as to the evolutionary reasons why plants in general retain more totipotency than animals. Plants do not move. They cannot escape misfortune. They must be able to regrow, to duck and weave, if some wind or herbivore chops them off in mid-flourish, or if they are forced to contort around some precocious rock. Animals do move. They have no such necessity. In addition, however, because they move they need a greater degree of internal coordination: their individual cells must be *more* differentiated, *more* committed to the needs of the organism as a whole.

Be that as it may; the fact is that at present, liver is liver, skin is skin, muscle is muscle, and you cannot take the nucleus from any such cell and coax its genes to produce anything other than what it has committed itself to be. At least, some animal cells show more pluripotency than others; but none, in a mammal at least, comes close to totipotency.

When that is possible, however (let's say in 100 years; which is a very long time in molecular biology), then it would be possible to take a cell from any tissue (perhaps); put it into the cytoplasm of an egg; and (probably employing some of the techniques discussed in this chapter) produce a new organism. Such a technology would take the Frozen Zoo into a quite new era. We would not need to store gametes or embryos; we could simply store tissues. Given that DNA is remarkably robust, we would not even need to employ remarkable methods of storage.

Another set of possibilities, which has a similar feel but is conceptually much simpler, is simply to keep samples of tissue (and the DNA that they contain) so that particular genes – some of which are bound to become extinct through genetic drift – could be reintroduced into future populations by genetic engineering. Such specific reintroduction is not possible now; but it is certainly possible (indeed easy) to store the necessary DNA with present technologies; and future conservation breeders may well find that whatever we store now is an invaluable resource. Dr Helen Stanley at London's Institute of Zoology is beginning such a 'bank'. It is all so conceptually (and technically) easy that it would be criminal of any society not to provide funds for expansion on a very large scale – and rapidly. A million pounds a year (the price of a warhead) could provide quarters for billions of genes from hundreds of thousands of individuals, from thousands of species.

This, however, is as far as it is worth speculating. Any watcher of the emerging technologies must feel a variety of emotions. One is excitement: it really is possible to multiply rare and beautiful antelope such as bongo, in the wombs of common eland. It is becoming possible to do the same for rare cats, and surely soon for primates.

Another emotion, though, is frustration. The places doing such work are few and far between: London, Cincinnati, San Diego. The number of scientists in the field in all the world seems barely to run beyond a few score. Much essential work is done by PhD students, who – through lack of funds – are often obliged to seek their living elsewhere after a single project. The financing in general is niggardly; it depends too heavily on the (albeit often prodigious) generosity of individuals, whereas a minute but committed proportion of the rich world's GNP could multiply present endeavours a hundred times (for we have come nowhere near to exhausting the *potential* scientific talent).

Correspondingly, the number of species to which it is so far possible to apply the kinds of technologies that are now reasonably advanced in cattle, humans, and mice – is small. We may rejoice at the birth of a bongo to an eland; but then reflect on how few times this has actually been done, and how many species of land vertebrate might in theory benefit from such interventions. We cannot give up hope: hope is necessary. But it is difficult sometimes to escape the feeling that the achievements at present (despite the sometimes superhuman efforts of the protagonists) fall short of what is truly required by several orders of magnitude. Sometimes it seems, too, as if the technologies themselves are 30 years behind the problem. If we could apply, now, the reproductive technologies that might be routinely available by 2200 AD – including IVF at will, embryo transfer between only vaguely related species, gamete and embryo storage, and cloning – then, with just a few million pounds, we might establish a truly comprehensive Frozen Zoo, with thousands of gene pools to pass on intact to our distant descendants, who perhaps might live in easier times and could restore the suspended creatures to a newfound wilderness. But by the time we get to 2200 AD, there might be very little left to put in the Frozen Zoo. There will certainly be far less than now.

The third emotion is a vague unease. I am sure that all the technologies discussed in this chapter are morally justified, given the context. It may not be aesthetically pleasing to make a chimaera between two species of rhinoceros (were such a possibility ever mooted; which at present it is not). But if, as a result of such a manoeuvre, the Javan or the Sumatran (or any other) were saved from extinction, then I would have no problem with it. Neither would I be too bothered if the decimated genome of some future population were eked out, with genes retained in storage from some present-day creature. "Is it better to do good on the Sabbath. . ."?

Yet – to go back to the opening fantasy of this chapter – I would be uneasy with a future technology that truly created new organisms; or even with one (only slightly less fantastic) that restored long-extinct creatures such as woolly mammoths (fond of them as I am) from DNA fragments in their frozen corpses. But it is reasonable, and necessary, to some extent to allow future generations to cross their own bridges. The task for the present is to do what we can to preserve what we have.

It is not sufficient, though, *simply* to conserve genes, or indeed to conserve the living animals that contain those genes. We must ensure that those animals are capable of living in the manner of their species; of behaving like rhinos, or gorillas, or whatever they are; and of holding their own in the wild. This is the subject of the next chapter.

SEVEN

THE WHOLE ANIMAL:
BEHAVIOUR CONSERVED

It is not enough, any more, simply to keep animals alive – or even alive and breeding. Curators acknowledge these days that 'state of mind' is as much a part of welfare as rude bodily health. And of course the ultimate aim is conservation, of which it is a condition that at some time the animals in captivity might return to the wild. Mere 'happiness' is not sufficient either, then. Animals in zoos must be encouraged to retain enough of their natural behaviour to make it possible for them to go back to the wilderness; or enough at least of their native wit to enable them to relearn the necessary skills.

Zoos, though, can be barren environments, and even at their best they rarely match the wilderness in complexity, and certainly not in un-predictability. The aim nowadays, then, is to create environments in zoos that maintain bodily health (as the wild emphatically may fail to do!) and yet keep the animals fulfilled and with the capacity to relearn the skills of the wild. This is achieved by the various techniques of behavioural enrichment.

With common sense alone and a little good luck, it is possible to keep an animal alive and reproducing. But to meet the needs of its psyche, and to keep it potentially fit for the wild, requires something more. It needs good science – knowledge of what the animal does in the wild, tempered by the insights derived in part from experimental psychology. But it is science of a modern and subtle kind: not 'pure', but admitting a fair and necessary measure of what might be called human sympathy.

AN ATTEMPT TO UNDERSTAND ANIMALS: ANTHROPOMORPHISM

The primitive and easy way to attempt to understand animals is simply to assume that they are like us: the approach called anthropomorphism. It has many manifestations. The portrayals of animals from Aesop to Edwin Landseer – cunning foxes and noble stags – are anthropomorph-ism of a kind; animals with human qualities, presented as symbols of those qualities. The animals of children's stories – Two Bad Mice and

Jeremy Fisher, Wind in the Willows, Rupert Bear, Donald Duck – *are* humans, in tweeds and spats and sailor suits. Because of the way they look and behave, however, they are presumed (as with Aesop and Landseer) to have some of the character of the animals whose physiognomies they have assumed: pompous Jeremy, bumbling Mole, irascible Donald. Poodles in diamante collars mincing to the limousine, are an expression of anthropomorphism; and so are rottweilers and pit bull terriers, wrenched through the park by muscly men with tattooed necks.

Anthropomorphism in these crude forms is woefully deficient. Clearly, at its worst, it degrades and diminishes animals: even Landseer's imperious lions in London's Trafalgar Square are in a sense insulting to lions, because they suggest that lions are admirable in so far as they resemble the Duke of Wellington, but *only* in so far as they resemble the Duke of Wellington. They are not admired in their own right, or on their own terms. To be sure, we should not be pompous about this: I have no quarrel with Beatrix Potter or with Rupert Bear. But we could argue none the less that our charming childhood companions do tend to get us off on the wrong attitudinal foot.

Crude anthropomorphism is also a poor guide to husbandry. Poodles, when let off the lead, are dogs like any other, chasing sheep and hanging on with their needle teeth. They are as capable as any other dog of going feral. It is hard to believe that they like sitting on silk cushions and walking like hackney horses (who are similar victims of human whim). Domestic animals suffer further from anthropomorphism because they are actually *bred* to fulfil human fantasies, and may be severely compromised as a result; pugs that cannot breathe, bulldogs that cannot give birth, big fierce dogs that by the standards of the wild are, quite simply, insane; for no wild wolf would act in the same way as Shakespeare's mastiffs, that 'run winking into the mouth of a Russian bear, and have their heads crushed like rotten apples' (*Henry V, III. vi*). Even when anthropomorphism is not so obviously grotesque it may be bad for the animal. Sir Peter Medawar, one of the great biologists of the 20th century, recalls how when he was a boy he tucked up a frog in a nice warm bed, which seemed to be a kind thing to do, but killed it.[1]

To be sure, common sense and experience avoids a lot of those traps – but even so, we cannot keep animals successfully just by assuming they are like us. We saw in chapter 3 how common-sense pairing of animals along human lines – young male, young female, privacy – does not necessarily (or even usually) lead to successful mating. Intellectually, too, such crude anthropomorphism is stifling. It pre-empts true understanding. As Darwin pointed out, no naturalist can hope to perceive what is happening in nature unless he or she already has a notion of what might be observed; the 'innocent eye' that Claude Monet tried to bring to bear is very difficult to achieve. But if the naturalist comes to the field with a preconceived notion that lions, say, are 'noble', then he simply will not believe the evidence of his own eyes if he sees them scavenging, or killing cubs, or lazing around while the females do the hunting. Natural history has suffered enormously from such purblindness, brought about by erroneous preconception. But natural history provides (or should provide) the data on which biological theory is founded.

So we need to move beyond the anthropomorphism of the storytellers and myth-makers. We need to employ the cooler objectivity of science.

THROUGH THE EYES OF THE SCIENTIST

Science in principle and at its best is a breath of fresh air. It sets out systematically to avoid all the kinds of traps that seekers after truth may fall into; and has come up with a *modus operandi* that undeniably offers the hope of progress – by which I mean progress in insight. Scientists and the 'natural philosophers' who preceded them have taken several thousand years to identify those traps, and we can never be sure that all have been uncovered. But there is cause for reasonable confidence.

The first requirement of good science – which is a general demand of all philosophy, and is not exclusive to science – is to avoid tautology, or circularity. It is tautological to suggest, for example, that lions are 'noble' because they possess some quality called 'nobility'. To be sure, this explanation would not be merely tautological if we brought some additional insight to bear; if, for example, we went on to show that nobility was some kind of substance, a hormone say, that produced specific effects.

Requirement two, is for repeatable observation, backed up by measurement. Observations are the raw material, the data; the bases of theories, the phenomena that need to be explained. But every judge knows how little store to place by the testimony even of the most fastidious eye-witness. It is so easy to be mistaken when you see something only once. Observations that can truly form the basis of scientific insight have to be as rock-solid as can be contrived. Solidity is achieved in part by measuring whatever is observed. In the field of animal behaviour this may mean, 'Under circumstance A, the animal did this 50 times, and that, 100 times; under circumstance B, it did this 100 times, and that 50 times.' Without such fastidiousness it is very difficult to get a handle on what an animal really does under different circumstances; and until we know what it really does, we do not even know what we have to explain.

The third requirement, which follows from the second, is – whenever possible – to control the circumstances. We cannot tell that there is a difference in behaviour between circumstance A and circumstance B unless we define carefully what the two circumstances are. Even in the laboratory this is difficult. We may do a hundred trials and conclude, 'If the animal sees a green light it does this, and if it sees a red one it does that.' But is the animal responding to the colour of the lights or their position? Many animals are colour-blind, after all. Does one of the lights buzz? Is that how it tells the difference? Most modern laboratory experiments involve a great deal more than a pair of coloured lights, and so leave enormous scope for ambiguity, which has to be overcome. In the field there is no control over circumstance, unless the observer chooses to carry out a field trial, and deliberately manipulates events. But control in the field is even harder than it is in the lab.

Largely because it is so difficult to keep track of all that goes on in the real world, and to control all the elements apart from the one under scrutiny, scientists try to break problems down into bite-sized tasks. This is one aspect of *reductionism*, an approach commonly thought to have been initiated by René Descartes. Mendel's investigations of heredity, which I discussed in chapter 4, provide a classic example of reductionism at work. To explain the inheritance of eye-colour in humans is impossibly difficult, as Darwin perceived – if you begin by looking at human eye-colour. But if you take a simpler problem – seed-colour in peas – you gain an insight into the more complex problem. Sir Peter Medawar summarised the point in the title of his book of 1967: 'Science is the art of the soluble'.[1]

A crucial insight into the true workings of science was summarised only in the middle of this century, by Sir Karl Popper. In a sense he was expanding the point that Darwin made about the need to have an idea in your head before you can observe, and which other philosophers had also made. Popper pointed out that scientists do not, as was commonly supposed, simply accumulate facts and wait for the explanations to fall out by themselves. Instead, he said, they constantly make guesses about the way the world works – hypotheses – and then put those hypotheses to the test. He also pointed out, however, that it is in theory impossible to *prove* any hypothesis. After all, whatever explanation you come up with, it could always be the case that something else entirely that you have not thought of is at work behind the scenes. What you can do in principle, said Popper, is *falsify* hypotheses: you can show they are not correct. It is the case, too, that not every hypothesis is of a kind that can be falsified; thus, for example, it would be extremely difficult to falsify the hypothesis that God exists. This does not mean that God does not exist; it merely means that the hypothesis that he does is outside the remit of science. Popper's point is, then, in summary, that science begins with hypotheses of a particular kind – the kind that can be subjected to experiment, and can in principle be falsified. The hypotheses that survive the best attempts to disprove them achieve the respectability of 'theories', and eventually come to be acknowledged as scientific 'truth'. 'Truth' that has been subject to rigorous investigation and survived is clearly not to be taken lightly, and is a lot better than a random invention that has not been put to the test. So science is far from trivial. But neither is science omniscient; it deals only in ideas of a particular kind, and always lives with uncertainties.

Related to the notion that you need to begin with a hypothesis, and related too to the general notion of reductionism, is the idea of the model. This implies that the scientist proceeds by comparing the thing that he seeks to understand, to something that he already understands; which then serves as the 'model'. It is intriguing, for example, the way in which our understanding of the human eye has largely followed behind our progress in technology. Thus, 300 years ago the eye was compared to the telescope. In the 19th century the 'model' for the eye was the camera. Now the retina itself is conceived as a combination of high-speed film and computer. Perhaps (as has sometimes been suggested) we will never have a machine that quite matches the eye; so perhaps the only proper 'model' for it is the eye itself. An intriguing notion; but in the meantime, in our attempts to understand, we must make do with the models we have.

Thomas Kuhn has expanded the notion of the model into that of the paradigm. He argues that all our understanding of the world and of the Universe around us is, at any particular stage in history, fitted into the framework of one particular and as far as possible coherent set of theories, explanations, myths, examples; a set that he collectively calls the 'paradigm'. The progress of science, he says, is not as smooth and inexorable as it seems. What happens, rather, is that as time passes, so new observations are made which do not quite fit into the existing paradigm. For a time, these new observations are simply explained away, or tucked out of sight. But eventually they became so overwhelming that the existing paradigm splits asunder, and a new and quite different view of the world is erected in its place, relatively rapidly, like an insect bursting its skin and emerging in a new form. As we will see, this view of scientific progress admirably explains the present shift in the 'scientific' view of animal psychology.

Finally, ever since the Middle Ages – since before the age of modern science – philosophers have acknowledged the need for parsimony. What this means is that you do not put forward a complicated explanation when a simple one will do; and indeed, you should in theory explore simple explanations exhaustively, before passing on to more complicated ones. The great arachnologist (spider specialist) W.M. Bristowe once offered a nice example of this. He was watching a small boy sailing a toy boat in the park (long before the days of remote control!). The boat scudded out to the middle, where the water was choppy. It reversed and returned to the shore, where it bobbed against the side near the small boy's feet. Imagine, said Bristowe, that he had been a good old-fashioned author of children's stories, like the excellent author of *Thomas the Tank Engine*. 'The boat', he would have said, 'bravely set out to cross the mighty pond. But when he got to the middle he saw how rough it was! "I don't like this one bit!" he cried, and headed back for the shore as fast as he could go. Bang, bang, bang, he buffeted the bank by his young master's feet. "Please lift me out!" he cried.'

It is fun, such story-telling (and in the hands of the Reverend Awdry *et al.* truly wonderful), but we can see that as a literal interpretation of events it is nonsense. We know that wooden boats do not think. The wind blew it out, changed, and blew it back again. The point is not simply that a thinking boat is a piece of anthropomorphism. The literary interpretation commits the more general philosophical sin. It brings ideas to bear that simply are superfluous. There is no need to assume that the boat has a brain. More to the point, if you do assume that it has a brain, and found a university department for the psychology of boats, you will first waste time and, second, fail to investigate the true cause. Only if explanations based on air-currents fail should you resort to something more elaborate.

These principles: repeatable observation; measurement and control; the testing of hypotheses; the creation of models; and the general principles of parsimony and the avoidance of tautology, have served science well. Without them, it would be an intellectual mess. Yet these approaches have their shortcomings; and in the field of animal psychology, these shortcomings have been exposed.

The trouble with too much rigour

To begin near the beginning. Measurement, control and repeatability have been the essence of solid progress. Yet they have drawbacks. To begin with, observations of animals in the field cannot be as controlled as rigour might demand; indeed, if we do try to control the conditions, then we influence the very thing we set out to measure. Observers overcome this lack of control in part by being astonishingly conscientious: recording virtually everything an animal does over representative (and generally huge) swathes of time and then applying statistical analysis to the results. This, too, is fine. It is a compromise, but it is the best that can be done. But some events are too rare to be 'significant' statistically, although they may be extremely significant in the lives of animals. Thus

Jane Goodall, who has studied chimpanzees for almost 30 years in what is now the Gombe National Park of Tanzania, has observed what looks very like warfare in chimpanzees (as we will discuss). I do not know how such an effectively 'one off' observation can be treated statistically, and how, therefore, it fits into the canon of scientific rigour. Perhaps it happens only once per century. But it provides an extraordinary insight nonetheless.

There is a drawback, too, in heeding too slavishly Medawar's admonition that 'science is the art of the soluble'. Of course that is true. It implies, though, that at any one time and in every field there will be a whole set of phenomena that we have not yet begun to investigate, largely because we are not yet equipped, perhaps intellectually and perhaps technically, to begin the critical investigations. The drawback lies in the complacent assumption that what we have found out so far is all that there is. We will see later that animals have suffered from this: from the assumption that existing knowledge provided an adequate foundation for husbandry. It was all too easily forgotten that the existing knowledge was limited – because it explained only those things that had been investigated, and the things that have been investigated are the simplest things, because simple things must be investigated before complicated things, because science is the art of the soluble.

Even the ancient rule of parsimony can be misleading. Of course it is silly to invoke complicated explanations gratuitously. We should remember, however, that we deal with simple explanations first not because simple explanations are necessarily correct, but because it is more convenient to begin with what is simple. Often, too, the simplest possible explanations seem to explain all that needs explaining. Even if they do, however, they may yet be deceptive. Take the case of the wooden boat. A description based on air-currents explains its behaviour very well. But such an explanation does not preclude the possibility that the boat may have a brain. The Rev. Awdry explanation could be the correct one. On balance, experience and common sense suggest it is better to opt for an explanation based on wind, and to stick to it until it really cannot stand up any more. Yet explanations based on wind would not explain why it is that human swimmers at the seaside generally return to the shore.

So – the fact that a simple explanation seems to work should not be taken to mean that the issue is closed. To lean too heavily on the principle of parsimony is itself a philosophical mistake. Come to think of it, even tautology can be useful. Perhaps there is a substance called nobility. It's worth checking out.

Science, then, is necessary: without good observation, measurement, the proposal of testable hypotheses and their rigorous examination, it is very hard indeed (effectively impossible) to improve on subjective impression and opinion; and subjective impression and opinion have demonstrably failed to provide a convincing understanding of animal behaviour and motivation. Science, in short, improves on common sense. Yet we can see on theoretical grounds alone that science, applied without the essential human restraint of common sense, can lead us astray. What, in practice, has science contributed to our understanding of animal behaviour, motivation, and predilection?

The rigorous view of animals

The early decades of the 20th century were a time of intellectual spring cleaning. In philosophy they saw the birth, *inter alia* of logical positivism; the attempt to strip down philosophy to a series of statements that could be considered to be unequivocally true, which alone (it was felt) were proper foundations for an understanding of the universe. Painters (*vide* the cubists) sought to discern and depict the essential forms which they felt underlay mere appearances. Architects (Loos, Le Corbusier, Gropius, van der Rohe) stripped away what they saw as the Gothic excesses of the 19th century and built again, like cubists, from irreducibly simple shapes.

So it was in science: in biology, and in the incorrigibly untidy field of animal psychology. To be sure, the animal psychologists of the early 20th century inherited a mess. From the previous three centuries had emerged a vast corpus of natural history; much of it excellent and painstaking (and some by true biologists, such as Darwin himself and Henri Fabre), but mostly a hotch-potch of travellers' tales. All was festooned in the trappings of anthropomorphism. George Romanes, friend of Charles Darwin, had tried to impose some evolutionary order, and constructed an elaborate hierarchy of animal behavour; fish, for example, he said, could 'recognise' but not calculate. All such observations are grist to the mill, and probably enhance understanding. But he too degenerated into romanticism and in any case the psychologists, as scientists, logical positivists *manqué*, needed something more substantial.

They found two solid platforms. The first was neurology: the study of nerves and how they work, which also had properly begun in the 19th century. Clearly (it had emerged) nerves work like electrical circuits (another 19th-century development). So here was a way of studying something real that was clearly relevant to animal behaviour. Indeed (an idea I harboured myself as a sixth-former) it might eventually be possible to explain the behaviour of every animal, from worms to Hamlet, in terms of electrical circuitry. The ways in which the circuits worked (though not necessarily the nerves themselves) could be studied in various ways. Notably, there were the studies of Ivan Pavlov on reflexes. His discovery of the conditioned reflex, in which a dog (say), could (unconsciously) learn to associate an otherwise neutral stimulus (like a bell or a light) with a stimulus that had innate meaning for it (such as the smell of meat) was particularly salutary: bell rings, dog salivates. When you put your mind to it, it is remarkable how much behaviour you can explain – or seem to be able to explain – in terms of conditioned reflex.

The second platform was behaviour itself. Nineteenth-century observers had tended to credit animals with various states of mind: happiness, thought, joy, sadness. They may indeed have such states of mind. But such states of mind cannot be measured; or at least, no one in the early 20th century had thought of any way of measuring them. What could not be measured could not be made a component of a solid, testable hypothesis. But if we want to make progress, the early-20th-century psychologists said, we must leave out things that we cannot get a handle on. What *can* be observed, and quantified, is behaviour. J.B. Watson in America is particularly credited with taking this purist (and necessary) line. He then, is generally seen as the true founder of 'behaviourism'.

Behaviourism reached its apotheosis in the middle of the century, at the hands of another American, B.F. Skinner. He improved on Pavlov with his own concept of 'operant conditioning', in which animals learned (effectively 'automatically') to do things that were suitably rewarded, and to avoid doing things that were punished. Thus, a pigeon that was rewarded with peanuts if it pecked a target after a light shone would quickly learn to peck the target whenever the light appeared. Operant conditioning combined with the conditioned reflex do indeed seem powerful; it is truly remarkable what you can *seem* to explain by applying them. More generally, Skinner continued to pursue the notion that all that was truly worth studying was behaviour; not (to be fair) because the states of mind that might underlie that behaviour did not exist, but simply because they were beyond investigation.

The behaviourists preferred to study animals in the laboratory, under strictly controlled conditions. Without controls, they felt, there could be no solid progress. There has also been a 20th-century traditon of observation in the field (or at least under more natural conditions). Acknowledged pioneers in this field – rigorous observation of more natural behaviour – are Konrad Lorenz of Austria and Niko Tinbergen of Britain. Lorenz often tended towards anthropomorphism, while Tinbergen in general veered more towards behaviourism. Neither tended to invoke explanations based simply on neurological circuitry – 'reflexes' – but they both invoked the notion that animals were driven by various innate 'drives' or 'instincts', which in modern parlance we might call 'programs'.

The achievements of the behaviourists and of the Lorenz–Tinbergen ethologists should not be underestimated. Their efforts did indeed bring us out of the murk of 19th-century myth-making and arm-waving. They did place the study of animal behaviour on a firm basis, in the laboratory and in the field. Essentially, what came out of both schools (though particularly the behaviourists) was that animals are machines that have been programmed by natural selection to respond 'automatically' to the stimuli of their environment; machines more complicated than clockwork toys, to be sure, but machines nonetheless.

Yet protagonists of both these schools fell into philosophical traps. The principal trap is to assume that because a simple explanation seems to work, it is therefore complete. It was necessary, in order to make intellectual progress, to concentrate upon the behaviour of animals, rather than to invoke the notion of 'state of mind'. But that does not mean that animals do not have states of mind. It merely means that states of mind had to be left out of the study for the benefit of the scientist. It was necessary, when investigating the behaviour of animals in the laboratory, to set them simple tasks in simple circumstances; because it is very hard to control experiments in which the task is complex and the options are various, and what cannot be controlled cannot give results that can be interpreted. But that did not mean that animals were incapable of solving complex tasks. It was possible, with ingenuity, apparently to explain much of what animals do by invoking simple mechanisms, of reflex and instinct; especially if (as in the laboratory) the animals were doing only simple things in simple circumstances. But that did not mean that reflexes, or instincts, provided the complete explanation.

In short, in order to explain the psychology of animals, early-20th-century scientists focused upon behaviour, because it was only their behaviour that could be observed and measured; and in order to explain behaviour, they erected a series of simple mechanical 'models' to explain what they saw. The models they erected were hardly more complicated than a clockwork toy. An ingenious piece of medieval machinery, with a suitable program, could do the things that animals were perceived to do.

In the last two decades, however, it has become apparent that animals can do a great deal more, and in a much greater variety of ways, than can easily be explained by such familiar mechanisms; and, indeed, that what they do depends very much on circumstances. The programmed machine model has become inadequate. At the same time, psychologists in the laboratory have learned to conduct experiments that are far more complicated than those of their behaviourist forebears (building on those forebears' experience) and yet are controlled; and so have learned to test animals under much more varied and taxing circumstances. Observations in the field, of an increasing range of animals, suggest that although animals may indeed conduct their lives according to programs of a kind (as indeed we may do ourselves), those programs are certainly not simple. They certainly consist of a great deal more than a string of reflexes. In fact, putting the whole thing together, it really is proving rather difficult to explain what animals do in a way that is not impossibly tedious, *without* invoking the once-forbidden notion of 'state of mind': confidence, contentedness, depression – all the myriad states in some form or another that have been described in humans by Shakespeare and Chekhov. In short, the 'paradigm', through which animals were perceived essentially as ingenious clockwork toys, has been broken.

BEYOND THE CLOCKWORK TOY

Some of the observations that have produced this paradigm shift were described at the Royal Society's meeting on 'Animal Intelligence' held in London in 1984.[2] There were reports of rats in mazes not of the traditional Y-shaped kind with cheese in one arm and a blip on the nose in the other, but in apparatus designed on the 'Trafalgar Square' model; a central space, with many arms, like the roads coming out of the real square in London. Food was placed in some arms and not in others, and in various trials and various combinations rats had to remember which arms contained what. The entire structure was rotated so that the rats could not take cues from the outside; smells were obliterated, so they could not simply leave a trail. Spacial memory was all that was being tested. (I am ashamed to admit that their ability in this, as described at the meeting, far exceeds my own. I used to work near the real Trafalgar Square, and I still have to think twice where all the roads lead.) But rats (and a number of other species that have now been tested) build accurate maps in their heads *en passant*. Owls clearly remember every branch in every tree – a three-dimensional map – and thread their way through the woods at night more by memory than by sight.

Marmosets, not generally considered the brightest of primates, showed a superb ability not simply to find rewards that were well hidden in complex environments, but also, apparently, to convey the where-abouts of the rewards to their fellows; and, indeed, they showed a willingness to do so. Pigeons are able to remember considerable orders of numbers and – which is far more significant – to put different parts of those strings of numbers into categories and hierarchies. Even goldfish – if experiments are suitably designed to allow for their physical limitations – can clearly learn, remember and discriminate far more keenly than most of us have been brought up to believe.

Behaviourism simply cannot explain these newly-appreciated abilities. Behaviourism supposes that animals simply respond in a prescribed fashion to an outside stimulus, like a fairground 'Try-your-strength' machine. Yet, as Herb Terrace of Columbia University, New York, commented after the 1984 Royal Society meeting, we can see now that 'the stimulus from the environment cannot fully specify or help us to understand what the animal is going to do' – provided, that is, the circumstances allow the animal some latitude. Furthermore, 'the animal itself is capable of generating stimuli'. And having generated its own stimuli, and received stimuli from outside, an animal can 'use them in a computational way, and given the results of that computation will make an appropriate choice'. This, quite simply, is a thinking process. We will come back to the issue of animal thought.

The laboratory and other highly controlled studies that were mostly described at the Royal Society meeting have been complemented since the 1960s by intensive studies of animals in the wild: studies of such duration, so closely observed and recorded, and carried out in such a representative range of circumstances, that they truly lent themselves to analysis and repetition, and indeed have all the qualities of substantive science. Among the first and most revelatory have been those of Jane Goodall, of chimpanzees, and the richness, variety, and complexity of chimp behaviour that her studies (and others') have revealed has surpassed all expectation. Chimpanzees have been shown to make tools – which before Jane Goodall's studies was thought to be the prerogative of humans. They have friendships, and personal dislikes; form alliances, run feuds. Their personalities are as variable as humans' – and individual personality largely determines survival: a group with a leader who is wise and generous fares better than one that is led by a tyrant. They have culture, such that different groups have different codes of behaviour and skills, which are learnt, and are passed from generation to generation.

At one point Jane Goodall's 'main study community' of chimps split into two groups, with a larger band in the north, and a smaller one in the south. For four years after the division, the two groups merely yelled abuse at each other across the no-man's land between them. Then the males of the larger group began a series of raids into the southerners' territory. When they found the southerners on their own, they attacked – 'in horrendously brutal fashion', she says.

Day-to-day skirmishes within groups usually last less than half a minute; but these were prolonged torments, spun out for ten or 20 minutes, and all the victims died of their wounds. These were not, indeed, simple quarrels. One group set out systematically – and that surely is the appropriate word – to wipe out another. I asked Jane Goodall if this could properly be said to be war. She did not actually say that all it lacked was generals with maps; but that was the gist.

Such behaviour as this, and all the rest that has now been catalogued in field and laboratory, cannot be explained by reference to clockwork toys. Indeed, as Pat Bateson, animal psychologist at Cambridge University, has put the matter: 'In some experiments, it is most helpful to begin with the assumption that the animal is like a human, until shown otherwise.' Have we come full circle, then? Are we back to anthropomorphism? Not quite, is the answer; or not, at least, in the crude forms that we have identified above.

Models are an essential tool of scientific progress: it is very difficult to see how we can get a handle on things that are not yet known, except by comparing and contrasting them with things that we know already. In explaining the behaviour of animals, the clockwork toy model has failed. We can elaborate the mechanical model: abandon clockwork in favour, say, of the computer. But animals do not behave like computers, either. Computers do what they are told, exactly, ploddingly, and one thing at a time; whereas animals at any one time may do one of many things, which are the outcome (we may say) of many different computations and impulses working simultaneously and in parallel. Yet we still need a model; something to compare the animal with. If toys will not do, what else can serve?

What is wrong with ourselves? To be sure, humans do not comprehensively understand humans. But we do have a unique insight into ourselves. We know the general dangers of anthropomorphism. But if we are critically anthropomorphic: if we sensibly compare and contrast what we know of ourselves with what can be perceived in other animals, surely that should give some worthwhile insights? Many others have gone down this path before. René Descartes in the 17th century was one. Among his observations was that animals do not think. He did not study animals, but he was known for his logic. And his logic ran: thought depends upon the use of language of a human kind – that is, on the use of words. Animals do not have a verbal language. Therefore, animals do not think. This seems to contrast with the comment of Herb Terrace at the Royal Society; for the processes Terrace was describing sound remarkably like thought. Descartes lived a long time ago, and he did not observe animals in the modern fashion, so we could just dismiss his comment. But logical arguments should not simply be swept aside, or they will come back to haunt us. How should he be answered?

Do animals think?

Descartes' assertion raises two obvious questions: first, is it true that animals do not have language that is comparable with humans'? And second, is language really necessary for thinking? For several decades, various American scientists have been exploring the first of these questions, through studies of chimpanzees. For some time, Herb Terrace was among them. To be sure, these scientists have not been addressing the general questions, 'Do animals have language?' and 'To what extent do they use that language in thinking?' But they have been asking, 'To what extent are the language abilities of human beings precedented in our closest living relatives?' Those studies are clearly pertinent to our theme.

Summarising, briefly and brutally, Alan and Trixie Gardner at the University of Nevada[3] have concluded, from very long studies on several chimps, that they can develop a considerable vocabulary; and that they can apparently manipulate items of vocabulary to frame brand-new thoughts.* Moja, for example, one of their chimps, rearranged 'listen' and 'drink' to signal 'Alka-Seltzer'.

One side issue, but one that adds excitement to such laboratory observations, is the work of Robert Seyfarth and his wife Dorothy Cheney in Africa. They have shown that vervet monkeys make different kinds of noise (alarm call) in response to different kinds of predator – and that other vervets, hearing each noise, respond accordingly. Thus if the alarm call signifies 'Leopard!', they climb into a tree. The 'Eagle!' call sends them diving for cover. 'If they hear 'Python!' (in vervet-speak) they stand upright, and look where the danger is coming from.

However, Seyfarth and Cheney have never suggested that these vervet calls are directly comparable with human language (though they do emphasise the richness of monkey social life). There are many who also doubt whether the albeit remarkable ability of chimps to remember and apparently to manipulate symbols is truly comparable with human language. The essence of the dispute lies in the philosophy of Noam Chomsky, of the Massachusetts Institute of Technology, who has become the doyen of 20th-century linguistic theory. The crux of human language, says Chomsky, is that it has syntax. Humans do not simply attach sounds to outside objects to form words; they do not simply have vocabulary. Neither do they simply join those sounds end to end, to attach to whole thoughts. Rather do they arrange the words in a hierarchy; classifying each one as a verb, a noun, or an adjective, and effortlessly tucking subsidiary clauses into main clauses. Toddlers do this, long before anyone teaches them the rules of grammar; or even that there is a thing called grammar. It is this ability instantly to classify and subclassify the component words, and not the acquisition of the words themselves, that truly constitutes language.

*I say 'items of vocabulary' in this ponderous way because chimps do not have a larynx that enables them to talk in the way that humans do – that is, to shape a huge variety of sounds – and so studies were generally done with sign language. But the principle is the same.

Herb Terrace worked with the Gardners for some years; but he is among the scientists who now reject the notion that chimps have language that is comparable with that of humans. He does so on two grounds. First, he says, simple rearrangement of 'listen' with 'drink' (and all the other comparable combinations) does not suggest a Chomsky-esque ability to put each word in its place in a hierarchy and to build a sentence accordingly. Chimps do not have syntax, in short; and syntax is what counts.

But – second point – Terrace also takes issue with Chomsky. He does not believe, as Chomsky maintains, that syntax is the fundamental quality of human language. The most basic characteristic, says Terrace, is the simple desire of humans – even of babies – simply to use words to draw attention to things in the outside world. Thus an infant will say 'cat!' not because he wants somebody to give him a cat, but simply to share the fact that he or she has observed a cat. Chimps, says Chomsky, never do this. If they signal 'drink' or 'listen–drink' it is because they want a drink. This is a remarkably simple observation, but the more you think about it, the more enlightening it becomes. This desire to share thoughts must surely precede the ability to do so, for how could natural selection have acted to refine that ability unless the ancestors of human beings were already putting such communicative abilities as they had to the test?

We can see, too, the evolutionary power of this desire simply to communicate facts about the outside world. Ancestral humans, as apes themselves, would already have been clever by the standards of their contemporaries. But once they began actively to try to communicate their thoughts, then they were able to 'put their heads together', and indeed to become collective thinkers. Books that refer to hundreds of previous authors, presenting thoughts that no single person could arrive at in a lifetime or even (probably) in the life of the Universe, show how potentially powerful such a predilection was; and suggests that natural selection would indeed have favoured its development, once it had something to work upon.

We may accept, then (I certainly accept) that apes do not pass the Chomsky–Terrace tests of human language. Unquestionably, however, they, and all other animals, communicate with each other. Without doubt, they attach sounds or signals to objects in the outside world, and those sounds and signals can be called vocabulary. Undoubtedly they communicate ideas through those sounds. We could put those three points together and conclude that animals do therefore, have language of a kind.

Undoubtedly, too, animals can in other contexts perform some of the mental skills that we associate with syntax. Many animals can arrange objects in hierarchies. Many social animals, too, do make gratuitous noises just to keep in touch with their fellows – you hear groups of squirrel monkeys chirping to each other high in the trees – and this could be seen as a precursor of the human desire to communicate wider thoughts.

But no animal except the human being puts all these abilities together, mentally manipulating symbols hierarchically, for the purposes of conveying thoughts. This coalescence of mental skills, together with our physical ability to express the symbols through a potential infinity of controlled sounds – in other words to speak – gives us a very significant edge indeed over all the rest of Creation. And though we may say that other animals do have language; and that different animals manifest at least some of the ingredients that we associate with language; nevertheless we are forced in the end to admit that only humans have human language, and that human language is in a class apart from all others. So let us ask the other of the questions that follow from Descartes' assertion. Is human, word-based language necessary for thought?

Consider Herb Terrace's summary of the Royal Society reports quoted above: that

> . . . the stimulus from the environment cannot fully specify or help us understand what an animal is going to do . . . that the animal itself is capable of generating stimuli . . . using them in a computational way, and given the results of that computation will make an appropriate choice.

What is this choosing and computing and use of memory if it is not 'thought'? What is thought, apart from the summation of these abilities? Unless we allow Descartes himself to be tautological, and to define 'thought' as 'that form of computation that is generated through and mediated by words', then we have to accept that animals do, indeed, think. As Herb Terrace says, 'The task now is to explain how animals think without *language*.'

We should not underestimate the things that other animals can do, or the ways in which they do them. Literally, and in all respects, they are not like us. But they are nothing like mechanical toys. We do indeed (if we keep in mind the caveats) see much more of the truth if we compare them to ourselves, than if we think of them as clockwork. There is one other characteristic of human beings which, in the past, we have chosen to deny in animals. We have emotions; and those emotions are extremely important to us, for they largely define our goals and determine our actions. Do animals have emotions, too?

Do animals have emotions?

We do not know whether animals have emotions exactly as we do. We certainly cannot divine what they are feeling. Furthermore, it is difficult to see how we could ever devise an experiment that would throw insight on what they are feeling. Science needs to work with ideas that can be tested. It is therefore difficult to see a scientific route into the problem of animal emotions; which is why the behaviourists abandoned the idea of animal emotion ('state of mind') in the first place.

A few common-sense observations seem in order, however. First, no human being can ever climb inside the skull or the skin of another human being; so none of us can tell precisely what another individual is thinking. Yet we do not doubt that other humans do feel something. All other observations suggest that other humans are, in all respects that are directly observable, similar to us: 'Prick us,' as Shylock said, 'do we not bleed?' We conclude that other people have emotions because we feel them ourselves, and it would be odd if in this one respect each of us were unique – and even odder if everyone else were lying about their own feelings. Furthermore, we tend to feel particular emotions under particular circumstances: fear, when faced by big dogs and runaway trucks, for example. When we feel particular emotions, we pull particular kinds of face, and signal with body language; and, lo and behold, other people, in comparable circumstances, pull the same faces and make the same bodily signals. The notion that other people feel emotions like ours is a hypothesis; but, taken all in all, it is far more plausible than the hypothesis that they do not.

I am never quite sure, either on common-sense grounds or in the light of evolutionary theory, why we should not apply the same logic to the question of emotions in animals. Anyone who has kept a dog knows when it *looks* happy, or *seems* to be afraid, or angry, or depressed, or self-satisfied, or any one of a group of comparable states. True, we cannot know that what a dog experiences subjectively as 'fear' is exactly the same emotion as we feel. But so what? It seems just as nonsensical to reject the idea that it feels *something* as it would to reject the idea that another human being feels something when his knees are visibly shaking and his face is pale and he chatters and points and stares. It seems reasonable to conclude, too, from all external signs, that whatever a dog experiences as fear it is unpleasant, if taken beyond the stage of mere excitement (which is also perfectly discernible).

Dogs also seem to register fear under much the same circumstances that we would; when faced by another dog that looks tough and is registering what, even to us, looks very like aggression. In short, to reject the idea that animals do feel emotions seems positively perverse. Furthermore, anyone who spends time with any particular animal, or group of animals, knows that they give signals that suggest an enormous range of emotions; that indeed there is hardly any emotion we recognise in humans that cannot be ascribed at least to the brighter animal species, without fantasising or lapsing into anthropomorphism just for the sake of whimsy.

Still there are purist scientists who might object on grounds of parsimony; don't invoke unnecessary ideas, particularly ideas that are not directly testable, to explain what animals do. It is possible to explain why a dog runs away from another dog that bares its teeth, without assuming that it is feeling fear. But we have been through all this. Psychologists have spent decades trying to exlain the actions of animals in strictly parsimonious style, and have failed. Animals do things that escape simple mechanistic explanation. We are obliged to seek new models, willy-nilly. And the idea that an animal is in fundamental respects like a human; that it does feel fear (or some equally unpleasant canine equivalent) when challenged, and that this is an important part of its motivation for running away is, so all the evidence suggests, a better hypothesis than the one which says it feels nothing. So it is silly to say 'a fox is cunning' or 'a lion is noble'. But it is not silly to suggest that each in its own way may experience a gamut of emotions that are at least analagous with our own. In fact, that seems an eminently plausible hypothesis; one that stands up to the albeit limited tests that *can* be done. Thus, to ascribe emotions to animals, provided this is done with proper regard to what can actually be observed, is now perceived, at last, to be not only reasonable common sense, but also good science.

Two final points, however. First, the modern realisation that animals are thinking and emotional beings does indeed suggest that they should be treated in a humanitarian fashion; that they must not be regarded as machines, that simply need to be oiled and garaged. We must not forget, however – for their sake – that the anthropomorphic caveats still apply. Each species (and each individual within each species) sees the world through its own eyes. Its perceptions are not the same as ours. Components of the environment that we do not perceive at all (such as smells and textures), or we do not think are important, may be of critical signifance to an animal. We should credit animals with emotions; but we should not lazily assume that all their emotions are the same as ours, or are evoked by the same causes. In short, we should respect animals as sentient individuals; but it remains incumbent upon us to explore the differences between them and us, to ensure that we supply them with what *they* perceive they need.

Second, however, we must accept that animal husbandry is very like gardening, or clinical medicine; that is, it must lean as heavily as possible upon science, because the information supplied by science is of a solid kind, even if it is not unequivocal; and because science does improve on common sense, and can supply quite new insights. But keepers, like gardeners and physicians, can never have perfect knowledge; and yet, they have to be prepared to tackle problems. In short, keepers and gardeners and physicians must be prepared to act in the absence of perfect knowledge, and indeed as we have seen must apply those intangible qualities known as 'instinct', and 'green fingers', and 'tender loving care'.

This is the general notion that lies behind the emerging field of behavioural enrichment. In the words of Hal Markowitz, now Professor of Biological Science at San Francisco State University, and author of the classic *Behavioral Enrichment in the Zoo:*

> We are emphatically confronted with the proposition that other animals besides ourselves like to do things, to see things change because of their efforts, to enjoy the pride of gathering their own food or drink, and to have some control over their lives. This is what behavioural enrichment is all about.[4]

BEHAVIOURAL ENRICHMENT

The philosophy of behavioural enrichment is still evolving – the shift in attitude to animals, which makes it now seem so obvious that they should not be kept in boxes with bars. The science is still evolving too: the behavioural science that is revealing what it is that each species really needs; and the veterinary science that allows curators to move in confidence from hyper-hygienic concrete floors and ceramic walls, to 'softer' earth and trees. But the common-sensical idea that animals need more than boxes with bars has been around for a very long time.

Outstanding and seminal in the modern age was Carl Hagenbeck, who founded Hamburg Zoo. It does seem that his chief concern was with spectacle; and as Jeremy Cherfas recalls in *Zoo 2000*,[5] he at one time ran a human zoo, where Europeans could stare at Nubians, Laplanders, and Buddhist priests. But he also cared about the welfare of his animals, and the naturalistic settings he provided for his animals, with a reasonable amount of space, and various, natural materials, ought, at least in general, to be an improvement on the simple cage.

Many 20th-century scientists, too, have concerned themselves directly with welfare. Thus the great American primatologist Robert Yerkes wrote in 1925 that 'The greatest possibility for improvement in our provision for captive primates lies in the invention and installation of apparatus which can be used for play or work.' Note the date: it was written at the time that Pavlov was working on reflexes in dogs. But note the language – 'work' and 'play' – which, by the 1950s, had come to seem highly anthropomorphic. Modern scientists who are involved in behavioural enrichment, such as Hal Markowitz in North America and David Shepherdson in England, quote Yerkes as a founding father (although the British in general are not quite so keen on the puritanical concept of 'work').

Hagenbeck and Yerkes represent (and largely founded) two quite different traditions of behavioural enrichment. Hagenbeck and his successors aimed primarily to create beguiling landscapes for the delectation of the visitors, with welfare being important but none the less a side-effect. Yerkes was a scientist, who had no paying visitors. He was interested first and foremost in the animals' mental well-being – partly for humanitarian reasons, and partly because he wanted to study ape behaviour, which is liable to be distorted in animals who are bored and neurotic. But note that he spoke of 'apparatus', which should be 'invented' and 'installed'. Appearance was not his concern. A gymnasium; carboard boxes; machines; anything that the animals responded to in a positive way was fine.

It is important, both for curators and for serious critics of zoos, not to confuse the two traditions. To a considerable extent they overlap, and complement each other. Savannah animals like to be in a Hagenbeck-style landscape. A tree is excellent landscape, but is also excellent 'apparatus' for an arboreal monkey: the best there is. But there can be conflict, too. Some animals that are kept in open landscape to give the visitors a good view prefer to be hidden in the bushes. John Aspinall in England is one of the directors who acknowledge this; his brown hyaenas at Howlett's in Kent are rarely glimpsed among the bushes that fill their enclosures. Nowadays, too, it is possible to create wonderful illusions of openness and naturalness simply by stage scenery: painted backdrops, trees of fibreglass and steel, potted plants in the background but out of reach of the animals. If such 'naturalistic' settings are designed with the needs of the animals in mind, so that there are plenty of different things for them to do and they can exercise choice – then fine. But it is all too easy to create a space that looks wonderful, and yet gives the animal no more freedom of expression than it would have in a simple box. We must always ask, too, whether artificial materials such as fibreglass really 'mean' the same to the animal as they do to the visitor. Human beings, for the most part, merely look at trees; but animals may climb them, smell them, feel them, bore into their trunks or probe beneath their bark. A fibre-glass tree, to some species at least, may be of no more interest than a concrete post. Neither should we accept natural Hagenbeck landscapes uncritically. Natural trees are excellent; but are of limited use to the animals if (as is often the case) there is an electric barrier to prevent them from climbing.

In general, it seems that the ideal must be to combine the best of the two traditions. Natural landscape on its own can be less stimulating than it seems and is also highly vulnerable: animals do tend to destroy trees, if they are in contact with the same trees day after day. But artificial environments – gymnasia, interactive machines – have to be very ingenious, complex, and expensive (at least in keeper time) if they are to compete with the sheer exuberance of nature. Trees, earth, sand and water are infinitely variable.

I said at the beginning of this chapter that behavioural enrichment (whether based on apparatus or natural landscapes) has two main aims: welfare, and the preparation of animals for the wild. Let us look at each of these in turn.

Rich environments for happy animals

In the late 1980s it was Dr David Shepherdson's job at London Zoo to improve the lives of London's animals and to develop the science of behavioural enrichment for the benefit of animals in zoos the world over. Accordingly, he approached the subject systematically. The first question to ask, he says, is how to tell whether an animal is happy or not: whether its environment needs improving; and whether the animal feels better after its environment has been altered. This question can be answered in two main ways.

The first is by physiological means: to measure the hormones that we know (not least from human studies) are associated with stress, such as the steroid hormones produced by the cortex of the adrenal glands. The advantage of this approach is its objectivity. Hormones are chemicals, and can be unequivocally quantified. The obvious problem is to translate hormone level into state of mind. For example, stresses are not necessarily unpleasant: excitement is stress of a kind, which it could be counter-productive to reduce.

The second approach is to observe the animal's behaviour: and here, says Shepherdson, there are again two possibilities. One is to observe how the animal behaves in situations that you know are unpleasant for it – when it is in pain or is attacked; for these reveal at least some of the signs that denote known (or eminently guessable) states of mind. The second, more general approach is to observe the animal's behaviour in the wild, and then see how closely it is matched by behaviour in the zoo. Of course, not everything that happens to an animal in the wild is 'good'; the wild is full of rivals, predators, and parasites. But although zoo curators might strive to remove the dangers and the diseases, they can hardly hope in general to improve on nature; and most would agree with Hediger's comment in *Man and Animal in the Zoo* that 'the standard by which a zoo animal is judged should be according to the life it leads in the wild.'[7]

In short, one eminently sensible way to assess the quality of the zoo environment is to compare the animal's behaviour in the zoo with the same species in the wild; and where possible, to quantify the comparisons. Three questions must be asked. To what extent are the wild and captive behaviour similar? (The more alike they are the better.) To what extent is the behaviour in zoos abnormal? Third, if the behaviour is abnormal, is it because the zoo environment provides some extra stimulus that is not experienced in the wild, or because it lacks certain elements, and leaves the animal deprived?

To take the first of these questions first: what exactly is it that animals do in the wild? What is the yardstick by which zoos judge themselves? Different species differ enormously in detail, and very few have been studied in great depth. But research so far shows some highly instructive patterns. Most wild animals spend an enormous amount of time resting. The outstanding exceptions are small warm-blooded animals such as shrews and song-birds, especially those with young, that must consume or gather their own weight of food (or more) each day. But ungulates (hoofed animals) probably spend up to 50 per cent of the time sleeping or dozing. Big cats, which do not need to eat every day and might in any case sit near a large kill for several days, may sleep or doze in the wild for up to 20 hours a day. Gorillas spend hours sitting around. And so on.

Wild animals also commonly spend a huge proportion of their time foraging and eating. Wild ungulates forage for up to 12 hours out of the 24. Mountain gorillas, with a more complex, forest environment to browse among, spend even more time in foraging; perhaps up to 45 per cent of the day. Bushbabies spend eight to 12 hours a day foraging.

Social animals spend a long time socialising: grooming, playing, courting, establishing their place in the hierarchy. Even when they are not devoting their time to this, they are generally interacting with their fellows in some way. Thus squirrel monkeys constantly call to each other as they forage in the trees. Herd-living ungulates prefer to be within sight of the rest. For social animals, indeed, curators agree that companions are the most important form of 'enrichment' that can be provided. But the groups do have to be appropriately structured for the species, and each individual has to be able to exercise options. If there is one thing worse for a social animal than a life of solitary, it is one – usually short – of constant harassment by ill-chosen, bullying cage-mates from whom there is no escape.

Finally, each such behaviour is underpinned by general activities: use of the brain (for animals do think, as Herb Terrace says) and of the limbs. Animals might in theory 'work out' in captivity, just as a human might do, and in the same ways: by solving puzzles that are in some way related to its activities in the wild; and just by staying physically active (for the same proportion of time as they would in the wild). In general, though, we must expect some differences between behaviour in captivity and in the wild. 'The wild' is not a single, uniform place; and each animal adapts to particular circumstances. So we cannot assume that every behaviour perceived in the wild is typical of its behaviour in all wild circumstances; and neither should we assume uncritically that any deviation from wild behaviour in captivity is necessarily bad. However, if the animal behaves exactly as it would do in the wild, then we can reasonably conclude that it is not unhappy. That is not a sure-fire assumption. But it is at least reasonable.

What, then, of the second general question? How can we recognise behaviour in captivity that is genuinely 'abnormal'? Excessive lethargy is one sign: unhappy animals may sleep even more than in the wild. Sometimes they overgroom: perhaps plucking fur and feathers until they are bald. Sometimes they are over-aggressive. Sometimes they overeat (obesity in humans too can be a sign of depression!) and sometimes they undereat. Unhappy animals often fail to breed (though, again, breeding is not by itself a certain sign of mental well-being). Many animals in the wild eat their own faeces (coprophagy) as a way of supplementing their diet, because excreta are full of bacteria and, in some species, contain a lot of largely undigested food, and can be highly nutritious; rabbits do this as a matter of course, and gorillas in the wild indulge in coprophagy now and again. But in captivity animals may do this to excess.

However, the most conspicuous form of apparently abnormal behaviour is *stereotypy*. The animal simply repeats the same action over and over again: perhaps pacing, perhaps head-bobbing, perhaps rocking from foot to foot, perhaps walking around in circles, perhaps plucking its own or a companion's fur or feathers. This looks appalling, and indeed may be; it very reasonably arouses public concern; it is seen in some animals more than others (and polar bears seem particularly prone); and it has invited some vituperative headlines of the 'Mad Bears' variety.

The special problem of stereotypy

It is not easy to define stereotypy. Clearly, though, it has connotations of repetitiveness and of abnormality. It does not seem to be the kind of thing that animals do in the wild. It does not seem to serve any 'purpose'. Empathy with the animal suggests, at the least, boredom and frustration. If we truly want to be helpful, however, we should approach all problems with a clear head. On the whole, parallels with stereotypic behaviour in mentally disturbed children, which some people have made, do not seem to be justified, whatever the superficial similarities. We know, for example, that most animals that are stereotyping can be cured or at least greatly improved if their environment improves – which suggests that in most cases the animal retains its capacity for normal behaviour, which is not what is normally implied by 'mad'.

We should distinguish, too, between behaviour that is truly neurotic and self-destructive, and behaviour that could pass as exercise. Polar bears in the wild are tremendous walkers, and as Graham Law of Glasgow Zoo points out, they have wonderful navigational skills, which they use to maintain position from month to month even though the polar ice cap rotates. True, their enclosures are usually small, and often are particularly barren, in a misguided attempt to simulate the Artic. Pacing in part is probably an attempt to take exercise; not necessarily entirely different from a person's routine stroll around the park. Some animals which are fed at regular times, such as tigers, commonly start to pace before mealtimes. This can be reasonably construed as nothing worse than anticipation. If they stand on their hind feet at the end of each lap (which they frequently do) then this could be an attempt to look over the heads of the crowd.

We should ask, too, does the animal pace all day, or only for a short time? Did we just happen to catch it at a 'bad' moment? Then again, is the activity one that it would normally do in the wild (such as walking) or is there some added pecularity (such as head-rolling)? The latter, on the face of it, gives more cause for concern than the former. Is the stereotypic behaviour associated with a particular place, or time of day? Does it pace by the door where the food comes in at feeding time, for example? Then again, is the animal easily distracted from its behaviour, or does it resist all attempts at diversion? And does it stereotype at the expense of normal behaviour? Finally, we should know the history of the animal before we draw too many adverse conclusions. For example, some zoo animals have had a chequered past and if they develop some stereotypy early in life in some unsuitable circumstance, they may continue it in a new environment even though the new environment may be better.

None of these comments is intended to deny that stereotypy exists, or that in many cases it probably does connote unhappiness. Neither should we deny that it can be severe. Some zoo animals stereotype virtually all the time, perhaps just rolling their head and moving from foot to foot, and stopping only to sleep and feed. Of course this is a problem. It is always important, though, to define what the problem is, and indeed to make sure that it really exists, before trying to tackle it.

Stereotypy has been extensively researched: yet as David Shepherdson points out[8], we are far from understanding it. What is the cause? Suggestions range from under-arousal, alias sensory deprivation, to over-arousal, including fear. Internal conflict may play a part – should I do this, or this? This sense of conflict is compounded by the animal's lack of control over its own conditions, and hence over its own fate; Hal Markowitz emphasises that the sense of being 'in control' is vital for peace of mind. Perhaps all these play a part, in different animals at different times. They are not mutually exclusive: it is possible to be sensorily deprived and fearful at the same time.

For what purpose, though, do animals stereotype? What do they get out of it? What is its *function*? Here again, says Shepherdson, the picture is 'equally diverse and baffling'. The most widely accepted explanation is that it 'helps animals to cope with an adverse environment. It might do this by regulating the degree of arousal or stimulation, increasing the predictability of the environment or making the animal less aware of its adverse surroundings or condition.'

But although the animals might benefit in these ways, the mechanism by which they do so is unclear. Research in the past few years has revealed the existence of endorphins in the brains of animals (including humans). These hormones latch on to the receptors to which opiate drugs attach, and through which these drugs operate. It seems, indeed, that endorphins are the body's own in-built painkillers and mood-ameliorators; and that opium has the effects it does because it is a reasonable chemical mimic of the endorphins (probably quite by chance), and because it is therefore able to latch on to the receptors which originally evolved to capture the body's indigenous endorphins. Be that as it is may, it is now clear that exercise leads to release of endorphins, which serve to suppress the pain of exercise; and this release of endorphins may be the reason that some people become 'addicted' to jogging, and feel very twitchy if they are unable to indulge. Perhaps animals stereotype so as to dose themselves on endorphins. Injection of drugs that block the release of endorphins, and so would render stereotypic behaviour less effective, do indeed reduce stereotypy (but only temporarily). However, there are many ways to interpret such a result and the issue is not resolved.

Recent studies of stereotypy – mainly in intensively kept farm animals – suggest that it develops in two stages. First the animal may react against some disturbing feature of its environment – for example a pig may pull against a tether; but then it may begin to make the same movement even if the tether is not there. As time passes the movement becomes simpler and refined and is enacted spontaneously, without stimulation from the external environment. 'It seems likely that many of the stereotypies seen in zoos never get beyond the first phase,' says Shepherdson. 'The pacing often seen prior to feeding comes to mind.'

When stereotypy does get beyond the first phase, the sequence of events in a zoo may run roughly as follows. An animal may move to the back of its cage to get away from visitors, finds that it cannot get any further, and begins to pace to reduce its agitation. Then, as time passes, it simplifies the movement and produces it even when there is nothing to escape from. In general, the kinds of changes that are commonly made in the cause of behavioural enrichment do tend to reduce stereotypy. If an animal has to forage for its food; if the food comes at different times of day; then it is not so likely to pace by the door where the food comes in (because it does not come in by the same route at the same time each day). Often, though, stereotyping animals give the impression that they just want to get out. Again, it seems naive to assume that most animals have a general conception of 'freedom'. They do not necessarily want to escape from comfortable environments. If they do want to escape, the task is to find the specific reasons. Here again, says Shepherdson, there are three main possibilities.

First, the animal may simply want to get away from a cage-mate that it does not like or is disliked by, or from a particular keeper, or from the visitors. 'Size of enclosure is of relevance. If an animal cannot maintain its flight distance from visitors or staff, then it will attempt to escape.' The second category is when the animal wants something, or to do something, that it cannot do in its enclosure: find a mate, or food; find a place that is warmer or cooler; a place to rest in comfort; to dig a hole or build a nest. 'In the wild', says Shepherdson, 'this would be expressed as exploratory behaviour.'

The third main possibility is when the animal in its enclosure sees something outside that it wants but cannot get to: prey; a potential mate; a competitor; a keeper coming with food. Or it simply wants to explore the area that it can observe. Note, though, that the views of the surroundings that Glasgow and Edinburgh take pains to provide for their carnivores and primates seem to alleviate frustration rather than promote it: and Apeldoorn's male gorillas like to be able to keep their rivals in view.

Behavioural enrichment is not of course concerned only with the alleviation of stereotypy; but reduction of stereotypy is in general taken as a sign that the enrichment is doing good. In practice, enrichment has been pursued through two main lines – which again should be seen to be overlapping and complementary. The first (in the style of Hagenbeck) is to identify those elements of the wild environment that the wild animal truly responds to, and recreate those elements in the enclosure. The second approach (shades of Yerkes) is associated largely with the name of Hal Markowitz: to promote general alertness and stimulate activity (of a kind that respects the animal's own biology) by the use of interactive machines. I will discuss this latter approach first.

Machines for working out

In the 1970s Hal Markowitz worked at Portland Zoo, Oregon, with the task of improving the animals' lot. He and his colleagues first began work with the Lar (white-handed) gibbon *Hylobates lar* because 'the zoo's existing protocol seemed especially inappropriate for the species'.[9] In the wild, gibbons live in monogamous pairs that may extend to 40 hectares. But at Portland, the gibbons had a cage of the kind that then was customary: plain walls and floor, a few steel pipes to swing on, and a rope. And although gibbons in the wild spend much of their time high in the forest of South-east Asia, Portland's gibbons, like most gibbons in zoos at that time, spent most of their time on flat surfaces, and indeed on the concrete floor where their food was delivered. The ideal – to recreate a wild habitat – was not an option.

The cage had four occupants: Mama, the mother, with her juvenile offspring Harvey and Kahil, and her infant, Squirt. The only reasonably active individual was Harvey, who went through a swinging routine for visitors. He used to slap the walls as he raced round the cage and was known (what else?) as 'Harvey Wallbanger'. The other three did nothing much except watch Harvey, and descend once a day to eat the food that came to them through a chute.

Markowitz and his colleagues installed an apparatus not dissimilar in general feel to the kind that the behaviourists favoured, but with a different intent: to give the animals something to occupy them. It was based on food, and it was high in the cage: after all, wild gibbons spend about 40 per cent of their time foraging, and they do this 'on the run', swinging (brachiating) high in the trees. Basically (though the design was modified as time went on) the animals could press a lever when a light went on at one side of the cage (near the top); then go to the other side, about eight metres away, to collect their reward. They all learnt quickly.

Early work soon confirmed what others had observed elsewhere; that if animals are given a choice – food on a plate, or food they have to work for – then they prefer to work for at least a part of their food. It soon became clear, too, that the animals' responses even to this simple apparatus were rich. For example, Harvey was happy to press the lever to release food on the other side of the cage when it was Mama who grabbed the reward, but would not perform for his siblings. The scientists used the apparatus, too, to involve the visitors, who were invited to insert ten cents to turn the light on, and start a trial. A notice made clear, however, that 'Animals are not machines and the gibbons may choose not to respond when the light is turned on.' None the less, the public's dimes added up to $3,000 in the first year, which went to research.

From these early days Markowitz and his associates moved in many directions, but always with the same general theme: apparatus to stretch the animal's minds and exercise their muscles. A family of diana monkeys from West Africa (*Cercopithecus diana*) learned (or at least the young ones did) to obtain a supply of plastic tokens by pulling on a chain (in response to a light) and then to trade these tokens for food. Again, extra richness of behaviour emerged. An older monkey refused (or was unable) to learn the trick herself but was certainly able to steal food from a younger one who had learnt. But the young one in turn learned to palm the tokens like a conjuror, leaving the old one in confusion. Neither was it always the obvious geniuses of the animal kingdom who learnt the tricks best. In one underwater variant of the gibbon apparatus seals quickly learnt what to do while dolphins, allegedly nature's Einsteins, looked on. Ostriches, traditional symbols of stupidity, showed they could manipulate apparatus designed to test their wits in ways that the experimenters had not thought of. This was par for the course: the animals usually thought of manoeuvres that had not been anticipated.

Such machines could clearly be of benefit to carnivores: for the principal 'foraging' activity of carnivores – hunting – is generally denied to them in zoos. Instead, for example, servals were given the opportunity to leap at swinging bars, which released pellets of chow; servals in the wild specialise in flushing birds and taking them in flight. Fibreglass marmots like the electric hares at greyhound races gave pumas something to chase, for rewards of meat.

There are snags in this approach. For example, cats in the wild do not hunt very much: but pumas in one zoo were seen to 'hunt' artificial prey up to 200 times a day. The behaviour itself may have resembled that of the wild; but there was far too much of it. In general, critics felt that if Markowitz-style apparatus was employed without proper constraint, then it would simply substitute a new form of abnormal behaviour for whatever abnormalities had existed before. At worst, the animals would become behavioural junkies, like children hooked on fruit machines. Markowitz concedes that this can happen – but if it does, he says, the fault lies not with the basic philosophy, but in its unintelligent application. Well-designed apparatus can help to keep an animal in physical and mental trim; it can (and of course should) reflect those skills that an animal uses and needs in the wild; and it can certainly improve on the once traditional, sterile environment.

In general, then, state-of-the-art (early 1990s) behavioural enrichment certainly incorporates many ideas that have been inspired by Markowitz's 1970s work, and certainly partakes of the general notion that animals do like to be active (at least for part of the day) and do appreciate variety and challenge. Nowadays, though, the more general trend is towards what have been called 'soft' environments, which incorporate as much natural material as possible; and laboratory-style devices, if they are employed at all, are given a more natural mien. David Shepherdson at London, and Graham Law at Glasgow, are among the British who are pushing in this direction.

The softer touch: London

Hal Markowitz went to great lengths to keep down the cost of his apparatus. All zoos – even rich zoos – have more tasks to attend to than they have cash to spend, and the more that is spent on games for gibbons the less there is to place elsewhere. His first apparatus at Portland was made in part from bits of an old mechanical coffee grinder. But scavenging for parts takes time, and time is money too. And then of course, machines always go wrong. It is in their nature. When they do, nobody puts them right. That is in their nature, too. Besides, some of the successors to Markowitz's Heath-Robinson originals were nothing like so puritanical. Antwerp had a 'locust-gun', like a Gattling, which shot live insects in among the fennec foxes. Grand. But each chamber had to be individually loaded, one locust at a time.

At London, then, David Shepherdson has always sought simplicity. He acknowledges his debt to Hal Markowitz, but is also inclined to quote Armand Chamove, who with his colleagues at Sterling University in the early 1980s pointed out the behavioural advantages of woodchip litter for primates. Woodchips are irreducibly simple; but each individual chip is a potential toy and food that is buried among them takes ages to find. In the wild, monkeys and apes spend much of their time searching among foliage or leaf litter, depending upon species. In general, too, Shepherdson has focused (as Markowitz did) on trying to enrich behaviour by making the foraging more interesting; for this, after all, is what occupies the greater part of the waking hours of most wild animals. Only their social life is more important (for a social animal); but this is taken care of by supplying suitable companions, or allowing the animals to live in their customary extended family groups.

London is an urban zoo, with a lot of animals in a limited space. Its enclosures are not large by modern zoo standards. In general, though, they are now made more interesting with supplies of branches (for nothing is more varied or more variable than branches) and with simple apparatus that makes foraging more of a challenge, and was (Shepherdson points out) largely designed by keepers. The simple equivalent of the locust-gun is simply a tube with holes in, stuffed with newspaper, in the interstices of which lurk crickets. The tube is hung from the ceiling, and every now and again – at unpredictable intervals – a cricket falls through one of the holes. A variation at London is a clear-acrylic perforated tube filled with mealworms in sawdust attached to the roof of the meerkat cage. Again, they fall through the perforations from time to time.

Marmosets have forward-pointing teeth in the lower jaw which they use in the wild to bore holes in gum trees; and the gum forms a substantial part of their diet (which their guts are specially adapted to digest).

London is among many zoos these days that have gum dispensers, made from hollowed wooden tubes and containing gum arabic inside, which the marmosets have to bore into.

One of the many revelations from Jane Goodall's work in the Gombe in the 1960s was that chimpanzees do not merely use tools, but make them too. *Inter alia*, they strip the leaves from twigs to make probes; and use these probes to poke into termite mounds. Accordingly, many a zoo these days has an artificial termite mound, of fibreglass or cement, where chimps poke for tit-bits of many kinds. Intriguingly, gorillas and orangs enjoy this too – though neither has been observed to poke for termites, or do anything similar, in the wild. Gorillas indeed may be keener than chimps, because they are quieter in temperament, and have a longer attention span.

At London, Shepherdson has introduced a variation on the theme which is nothing like a termite mound to look at, but none the less is 'functionally equivalent'. It is a plastic pipe fixed just outside the cage and containing pieces of fruit, which chimps push along to the end with twigs that they modify themselves, and poke through small holes in the tube's sides. Other zoos have devised many variations on this theme.

Finally – a piece of behavioural enrichment that has to do not with foraging, but with social life. Gibbons in the wild stake their territorial claims by singing, usually in male and female duets. Lars sing for about 15 minutes at a time, sometimes only now and again (once in five days) but sometimes twice a day. The point of the song is to keep the different pairs apart. Yet the song is part of social life none the less. It seems that one of the effects of hearing a rival is to strengthen the bonds within each pair.

London, like many zoos, has only one pair of gibbons. David Shepherdson felt it might enrich their lives to hear the singing of others, so he fixed up a broadcasting system which plays the recorded song of wild lar gibbons through a 30-watt loudspeaker, from the top of the main primate house about 50 metres away and 10 metres from the ground. This was on an endless 60-second cassette and, he says, 'contained several sequences of alternating male and female calls of increasing intensity climaxing in a long screeching call from the female, the great call, and a warbling reply from the male'. This was played twice a day for 14 minutes at 11 am and 4 pm, with pauses of one to three minutes, which is roughly the pattern of singing in the wild. Beyond a doubt, the gibbons responded to the song. In the half-hour after it, they were significantly more active, and spent far more time at the top of their cage and in brachiating, than in the half-hour before. They did not grow bored with the singing, either; that is, they did not 'habituate' to it, and come to ignore it. Many months after installation they were still responding.

Big carnivores, though, pose special problems: because they are natural hunters; because they are potentially dangerous and difficult to handle; and because they are big, and anything done for them is liable to be expensive. But Glasgow Zoo's keeper of carnivores, Graham Law, and his colleagues have shown that much can be done for them as well.

Glasgow: big cats and small cats

It seemed a good idea to feed the cheetahs at Glasgow by throwing their meat over the fence of their enclosure – which must be more interesting for them than the usual boring bowls. Normally the cheetahs leapt and grabbed the meat but sometimes they missed, but if ever it landed on the ground they seemed bewildered. They would come back to the fence with a 'Where's it gone?' look on their faces. The Glasgow keepers noted, though (as others have done) that cheetahs sit for hours, staring into the distance; after all, they are daytime predators who hunt by sight, and they pick out their prey at a distance and run it down. Long vision is their forte. They do not have an outstanding sense of smell, or good close vision, and they miss things under their noses.

The first move, then – which has become a general feature of the husbandry at Glasgow, as well as at nearby Edinburgh – was to build a high platform, so that the cheetahs could look out over the nearby elephant house, to the cars on the motorway that runs through Glasgow, and to the deer next door. The arrangement works well. The cheetahs use their platforms all the time. In the wild (as field studies now show) cheetahs climb to the tops of termite mounds to defacate and look around. Glasgow's tigers, too, were given a large, high platform where they now spend much of their time. They have a view of the riding school that lies beyond the river.

The cheetahs and tigers at Glasgow have large enclosures but the leopards and small cats have smaller cages. To improve the general environment Law and his fellow keepers introduced rocks, branches, and woodshavings. But still the animals were fed only once a day, on meat that was easily bolted. The first change was to feed the cats three times a day. Then came games. The keepers suspended the leopard's food from a pick-axe handle hanging from the roof, putting it in a different place each day, just out of reach. Each mealtime the leopard has to work out how best to approach the problem; whether to leap from the ground, or from a branch, and if the latter, then which one. As Markowitz found, animals, when posed a problem, are ingenious. One of the leopard's solutions is to leap from the ground and set the handle swinging; then leap to a branch and catch the meat at the end of its swing.

For the ocelot the keepers tried a variation on the theme. A pole hung from the wire roof; at the top of the pole a tupperware box with the meat in, which the ocelots had to fish out from below, once they had climbed the pole. To make the whole thing more difficult, the pole was hung from a swivel. To be sure, the first ambitious design had to be modified, as the ocelot spun round 'at a hundred miles an hour', and landed in a dizzy heap. But the principle works, and the swivel has been replaced by a simple string. There are even easier devices, too; like hiding the food in a pile of sticks. In the wild, small cats fish in just such places for rodents. It all makes the day more interesting; the environment more complex and unpredictable, as it is in the wild. Pick-axe handles and piles of sticks are cheap. All it takes is keeper ingenuity, and time.

Glasgow's polar bears

Polar bears have been a huge challenge to zoos. They are, as Law says, 'easy to keep alive', and they breed reasonably well these days but they are among the most notorious of all stereotypers: pacing and head-rolling, rocking and head-bobbing are familiar and distressing sights. Even zoo enthusiasts have often doubted whether polar bears should be kept in captivity. What, though, are the problems? In the wild the polar bear is an active animal, migrating huge distances, hunting at sea. But this need not be a disadvantage. Other lively species, including primates, can do well in captivity.

It seems obvious, though that innately lively animals need stimulating environments, and of all zoo enclosures none is likely to be more barren than the polar bears'. This point is not that curators have a down on polar bears. But economy, logistics, the bears' physique, and lack of knowledge of their true requirements have conspired to give these animals a bad time. Polar bears, to begin with, are immensely strong, and are also the most carnivorous of bears. Security is at a premium. A cheap and easy way to ensure security is to build a deep pit with smooth concrete walls. That is generally the starting point.

Because they are carnivorous, too, polar bears tend to leave old bones around their dens. These, combined with the litter that is inevitably dropped and blown into a pit, can rapidly produce a scene of utmost squalor. Keepers need to lock the bears in their dens when the visitors are gone, to clean the outside pens. But this is difficult because the bears are reluctant to go inside – they do not like being locked in. If there are two bears they cannot be lured inside even with food, because they tend to go and fetch the food one at a time, and re-emerge to eat it. Largely to reduce the inevitable cleaning problem, then, zoos traditionally fed chopped meat and vitamin supplement, and perhaps whole chicks, but nothing that would leave litter. Hence, a boring diet, that the bears could wolf down in a few minutes. Finally, with limited knowledge of polar bears in the wild, but with a desire to create a 'natural' environment, zoos traditionally tried to imitate ice and sea with hills of concrete and, of course, a pool. Concrete is not a sympathetic material.

Graham Law and his colleagues inherited a traditional polar bear pit at Glasgow that had been built in 1960 and in its day was among the biggest in the world: an acre (0.4 hectares). Once it had six bears, but by the time Law came on the scene the zoo had only one male and two females, all too old to breed, and stereotyping. He and his colleagues set out to solve the problems systematically. First, they reasoned that the male bear stereotyped by day partly because he had nothing to do but wait for his evening meal. So – very simple! – they decided to feed him in the morning instead. 'It was like a miracle,' says Law. 'He would get all his scoff down his throat and then he would lie about and sunbathe, and then swim around in his pool – which was very normal behaviour!' So then, to make his life more interesting still, they began to feed him several times a day. This worked even better.

A second innovation was to put straw in the bears' dens to sleep on. It may seem astonishing that polar bears were not traditionally given straw; but then, they live in the Arctic, don't they? And there is no straw there. At first the bears dragged the straw into the open. But soon they realised it was for sleeping on, and comfortable; and this in large part solved the problem of getting them in at night. Thus it became easier to clean the enclosure, and so it became possible to feed them a more interesting diet. They were given whole cow legs, and learnt to crack open the bones for the marrow; and entire rabbits, which presented them with a challenge and gave them animal fibre. Dead fish were thrown into their pool, and live crabs.

Neither are polar bears quite so carnivorous as is generally supposed. In the wild, given the opportunity, they dive for kelp. In captivity, as several zoos have now shown, they appreciate all manner of vegetables and branches such as willow – which they like to be spiced with garlic. A further modification is to freeze the tit-bits, vegetable or animal, inside blocks of ice, and float them in the pool. The bears appreciate the challenge. Polar bears, too – as many zoos now have found – love toys. They like big bright things that float – such as traffic cones, and five- and ten-gallon plastic containers. At Glasgow they developed a special game with the latter: bringing them ashore and pummelling them with their front paws; crashing down on them with their whole weight. In the wild, they break open the ice-dens of seals in just such a way. Young polar bears in the wild perform handstands on the ice-dens to make up for their lack of weight.

Not every toy keeps its charm forever. The keepers must strive constantly to devise new distractions. This, combined with the multiple daily feeds, and the increased cleaning, adds considerably to the workload. But, says Law, 'That can hardly be grudged as the polar bears have themselves given all their time to the zoo and the zoo-going public.'

The overall truth – one illustrated by the work at Glasgow and like-minded zoos, and supported by evolutionary studies – is that polar bears are *not* hairy seals. First and foremost, they are *bears*. The palaeontological record shows that despite appearances, they probably evolved from brown bears only 70,000 or so years ago. True, they like water. True, they are supremely adapted to the Arctic. But they like to do the things that bears do: tear branches; dig holes; and even (as the female polar bear at Edinburgh Zoo showed when she was put into the brown bear enclosure), climb trees. Also, like many wild animals, they appreciate a good view. In short, the traditional concrete pits that were built in pale imitation of the North Pole are the very antithesis of a good polar bear environment. Polar bears like rocks for sunbathing and plenty of space to swim, and all the diversions and varieties of diet the keepers can think up; but they also appreciate hills and trees, and earth and grass to dig in. The enclosure should be big – we always should be thinking in terms of acres – and ideally, perhaps, only a part of it should be on public display. There should be fields and woods for the animals to retreat to. Variety, choice, a good view; these are the fundamental elements of husbandry for intelligent animals, and as we have seen, most animals (especially bears) are far more intelligent than they have generally been given credit for.

Glasgow, though, has not yet (1990) been able to build its ideal polar bear enclosure. It has been obliged simply to modify what was there already. For its Himalayan bears, however, it has been able to start from scratch. This, then, might truly represent the enclosure of the future.

Glasgow's Himalayan bears

In the mid-1980s Glasgow had no specific plans to acquire Himalayan bears. Bears in general were proving controversial in zoos; and Glasgow was still solving the problems of its polar bears. As Graham Law put the matter, 'Do we really want to get in this barrel of worms again?' But there was a park nearby that specialised in keeping bears; or at least it would have done, but the bears escaped every now and again and the local council withdrew their licence. Glasgow was asked to find homes for four of them, and were given some financial help to get started.

Although Glasgow is an urban zoo, it does have plenty of land. Item one, then, was to designate three acres to its newly acquired Himalayan bears, including a two-acre wooded hillside with a stream at the bottom. Around the entire enclosure runs a tall wire fence (deeply buried at the bottom) topped by a smooth barrier (which in places is made of toughened glass) to prevent them climbing out.

From the beginning, the idea was to enable the bears to spend their days as nearly as possible as they do in the wild. One thing they do is to make nests, known as enza. Because the hillside was wooded, there was plenty of material about, and places to make them; and the keepers added more nest materials as well. In practice, the bears spend a great deal of time making their enza, which are neat and elaborate. Visiting biologists have expressed amazement that the bears make such nests themselves, and that the keepers have no hand in them.

The bears have a custom-built house, both for nightly sleeping and for hibernation. They sleep in metal baskets – 'like Aunty Freda's hanging geraniums' – which Graham Law designed and which were made by a local blacksmith. They were expensive, and it was not certain that the bears would take to them. But they have. They line them with straw and sleep the nights and the winter through.

The bears too have a look-out platform, like the tigers and cheetahs. They too enjoy the view. As a bonus, the females have a better head for heights than the male, and can climb out of his reach. He can, after all, be aggressive. A huge woodpile on the flat part of the enclosure at the top of the hill gives them something to hide behind – or would, if they did not pinch so much of it for nest-making.

In the mornings, the keepers attach small pieces of meat to peg up the legs of the platform, which the bears climb for. Meat, fruit and vegetables are scattered throughout the three acres. Raisins are a great treat. Small though they are, hidden in the undergrowth, they do not escape the bears' investigations. Other food is hidden in the woodpile. Yet more is dropped down drain-pipes buried vertically in the ground, into which the bears must reach. There are parts of the enclosure that the bears do not generally make use of – which seems a waste. So the keepers tie small pieces of meat to a string and drag them into those corners to leave a scent trail (a ploy suggested by an animal psychologist). The bears follow such trails even though the piece of meat is small. Sometimes the meat is left at the end of the trail and sometimes not. The wild is unpredictable too: sometimes hunts are successful and sometimes not. All in all, foraging takes up much of the day, as it would in the wild.

The Glasgow keepers feel there is still much to be done. They have looked at the possibilities for scattering food throughout the wide enclosure with a cross-bow, or a clay-pigeon shooter; but have not yet come up with a workable method that does not involve them in trekking around. The stream at the bottom of the hill could become a source of fish, dead or alive. Again, the possibilities are endless; but already, the lives of the Glasgow bears do not seem very different from those of the wild. Visitors can watch them either from the top of the hill, through the fence: or gain an unimpeded view from the valley on the other side of the slope. I suggested that a small grandstand be built, so people could watch as if at a cricket match. Watching animals living natural lives can be an extremely agreeable pastime.

David Shepherdson's task at London was in general to improve the environment in existing enclosures. Glasgow is upgrading some, and creating others from scratch. But already there are a few zoos that have been designed entirely, from the beginning, around the behavioural needs of their animals. Two of these have already been mentioned several times in this book: the Monkey Sanctuary at Looe in Cornwall; and Apenheul, at Apeldoorn in Holland. These are eminently worthy of discussion.

England's Monkey Sanctuary and Holland's Apenheul

Cornwall's Monkey Sanctuary is devoted to one species: indeed to one subspecies of the woolly monkey, *Lagothrix lagotricha*, from South America. At the time of writing it contains only 20 individuals: a single population now into its fourth generation, with a dominant male Charlie, a number two male, Django, females, juveniles, and infants.

The sanctuary's founder, Leonard Williams, would not have liked it to be called a 'zoo'. When he began the sanctuary, zoos were not the places that some of them have become. Monkeys in those far-off days (was it really as recently as the 1960s?) could be freely exported (which Brazil now forbids) and equally freely imported, and as woollies are sociable, intelligent, and reasonably quiet by monkey standards, they were among the favourites. But most of those that were imported to Britain were kept in solitary as pets or as photographer's props; and it was these animals that tended to wind up in zoos. Zoos in turn tended to keep them in barren boxes, sometimes in pairs, where the males, with no proper group to lead, were inclined to bully their intended marriage partners horribly. Breeding was rare, and successful rearing even rarer.

Williams was accused by many professionals in those days of the cardinal sin of anthropomorphism, while he in turn railed against behaviourists, in conversation and in his books. He set out not to treat the monkeys as mini-humans, but as sentient and intelligent creatures with their own outlook on life that demanded respect. They had the run of his house, together with the wire enclosures that formed an extension to it. He brought the animals together in the same way that zoos did – 'rescuing' them from unsuitable homes. But they quickly formed a colony and bred almost from the beginning.

The sanctuary is in many ways remarkable. The 20 animals are cared for by no fewer than 15 keepers, all living in, as a commune; 10 of them resident, and five or so transient volunteers (including students of zoology and trainee keepers). But it is the points of husbandry that are of interest here, for they illustrate perfectly the modern principles of behavioural enrichment. First, the monkeys *are* a colony; as social animals, the most important things in each of their lives are their companions. Their feeding regime is a model of good sense. They are fed twice a day – more than they can eat, because otherwise captive animals eat too quickly, and the herbivorous species are liable to finish up with bloat (gas in the alimentary tract). The food (a wide variety of fruit and vegetables, for woollies are largely though not exclusively vegetarian) is only roughly cut up, so they spend a long time picking out and tearing off the best bits. The keepers also bring them regular piles of branches and wild herbs; they like sycamore, for example, not because the leaves are particularly pleasant, but because of its inevitable cargo of insects. The monkeys also browse and graze for themselves on the plants in their enclosures; and they have access to a mature tree where sometimes, in addition, they find birds' eggs.

The enclosures themselves are enclosed with wire roofs as well as walls, but they are spacious and full of interest. The keepers constantly observe the monkeys' needs, and continually add ropes and branches. Most of the time the enclosures are all linked together by runaways so that they form a continuous territory, which includes part of the house. But when the colony was somewhat smaller, one entire enclosure was usually kept closed to them; they went in only now and again as a 'treat', with a great sense of adventure each time, to find the vegetation well grown during their absence. Leonard Williams records that the dominant male would enter first; and, among other things, would break dead branches off the trees before the others were allowed on. (Whether this was to spare the others injury, or because he liked breaking branches, is open to debate.) Above all, though, the monkeys can always find alternative routes around the enclosures. This is partly for variety, and to give the animals choice; but partly so that they can, if they choose, avoid individuals they do not care for (for they certainly have their likes and dislikes), or give the alpha male a wide berth if it seems that he requires it. The point is that no animal need ever feel cornered by another, as so easily happens in traditional enclosures.

Of course the final test for any zoo animal and its keepers is whether it can return successfully to the wild. We will discuss specific training for the wild later. In the mean time we may recall from chapter 5 that two of the Monkey Sanctuary's young males are now (early 1991) returning to the Brazilian forest not simply to try their luck, but effectively to take charge of a group of young, wild-born monkeys. If they succeed (and there is every reason to suppose that they will) then this will be a triumph indeed for husbandry and for captive breeding.

In Britain there is now a law which forbids contact between visitors to zoos and non-domestic species. However, the law contains a clause which allows the Monkey Sanctuary to continue its traditional practice – of allowing some of the female monkeys to meet the visitors at the end of their visit, and even to sit on the visitors' laps. The choice is with the monkeys, however. They meet the visitors only when they feel like it.

Apenheul, in Apeldoorn, takes contact between animals and humans probably about as far as it can be taken. Most of its 18 species of primate (which include about 30 woollies) are able to make contact with the visitors as much and as often as they (the animals) choose. Only a few, such as the gorillas, are permanently sequestered, on their pair of islands. But if the animals want to stay away from the visitors they can, and if they ever feel they have had enough human society they can get away immediately. Everywhere there are escape routes; the squirrel monkeys descend from the trees to pick the visitors' pockets, but they are up and away as soon as they grow tired of the sport.

The design of Apenheul is a miracle of ingenuity. The bordering fence – again, a smooth barrier above wire – is tall, but well hidden in the trees. The zoo itself, a natural landscape, is divided into several discrete areas separated by moats; effectively, a series of islands. Species that get on well together are on the same island, and those that do not are separated. Likes and dislikes are unpredictable. Woolly monkeys get on very badly with spider monkeys, though they live in the same part of the world and probably encounter each other in the forest; but they get on well with lemurs, which live two continents away in Madagascar. Visitors pass freely over the bridges between the islands, but the monkeys do not. Why? Because in the winter, when there are no visitors, the dominant males of each species are taught not to cross the bridges, and what the dominant males will not do, the subordinates will not do either. Learning is achieved by a small electric shock, delivered by a 'hot wire'. This is turned off in summer – but by then, once the monkeys know what they should not do, it is not needed.

There are dozens of little touches of good husbandry. Of course the monkeys forage for much of their own food: the squirrel monkeys form a single troop of 100 or more individuals (as they do in the wild) and flock through the trees like birds, twittering to each other to keep in touch, summoning their companions to caches of fruit and insects. As Erich Mager of Apenheul says, 'visitors should have a crick in the neck by the end of the day'. The food the monkeys can forage for includes herbs – parsley, coriander, mint, and so on – which grow under the wire mesh, so that the monkeys can pick the outer leaves without uprooting the plants. The point of this is that wild primates (and, probably, many other wild animals too – including cats and dogs, of course) seek out particular herbs when they are ill, and self-dose. Observations at Apenheul suggest that this is precisely what happens: for example, monkeys that visit the herbs tend to be ones with wet stools, suggesting some digestive upset. The subject of animal pharmacology is potentially huge (for most wild plants are pharmacologically active to some extent) and is only now being broached.

Wim Mager (Erich's brother), who founded Apenheul in the early 1970s, wanted at first to allow the gorillas, too, to have contact with the visitors. In the end, discretion prevailed, and they were given a large island instead (which they share with ground-living savannah monkeys that are extremely fast on their feet). The gorilla colony grew, and it became necessary to divide it. The Edgar Rice Burroughs myth that gorillas are neanderthal hunters has given way in recent years to the belief that they are endlessly amiable herbivorous giants – but that is not quite true either. Each gorilla colony is typically led by a big silverback male, and dominant silverback males do not welcome others in their vicinity. The problem was, then, to divide the gorilla island in a way that would minimise friction between the two new groups.

The problem was far from easy. The obvious solution was to build a tall wall. But this, first, would be potentially ugly (as well as expensive); and, second, was forbidden by the local council; and, third, would leave the alpha males within smell and sound of each other, but out of vision: a possible source of frustration and stress. The second notion was simply to run a moat through the middle. However, this would leave too much visual contact between the rival males; and in their excitement, it seemed quite likely that one or other would drown in its efforts to get at the other.

The Magers and the keepers between them finally came up with a marvellously ingenious solution. The two gorilla colonies are indeed separated by a moat. But on either side of the moat is a steep dyke, with vegetation on the banks, and an electric fence running through the top. To be sure, an angry and frustrated silverback could easily crash through such a barrier. But on each island, standing a dozen or so metres back from the dyke, is a hill. Each alpha gorilla can stand on its own hill and shout abuse at the other. However, when they want to approach each other, they have to come down the hill, and so lose visual contact, which reduces the stimulus. To re-establish visual contact, they have to move away from each other, to re-ascend the hill.

That was the theory. The dykes and the hills were built at great expense. Nobody knew whether the arrangement would work. In fact it worked like a dream. Each morning, the two males climb to the top of their hills, flex their muscles, fluff out their hair, roar, and generally make themselves fierce and beautiful. Honour satisfied on both sides, they come down again and spend the rest of the day in peace. That is applied psychology – and proves that psychology can work!

One further refinement. It seemed important that the gorillas should know that there was indeed a moat the other side of the dyke that runs alongside; and that they should know this without having to climb the dyke. So each colony of gorillas has a beach, at one end of its dyke. The two beaches are at opposite ends of the dividing moat, however, and the moat is curved, so that when both groups are on their beaches, they are out of sight of each other.

Apenheul demonstrates an important principle that we will return to in chapter 8: that animals can adjust their behaviour to people without unduly compromising their own biology; and that people at large are perfectly able to respect the needs of animals. The next conceptual step beyond Apenheul is to keep animals in a national park – living entirely wild (although managed) but observable from hides. Such places already exist, of course, as Jeremy Cherfas describes in *Zoo 2000*. But places that keep animals at a distance from their native habitat, like Apenheul and indeed like good zoos everywhere, have a vital role to play as well. And if all zoos follow the modern paths of behavioural enrichment, even if they feel they must fall short of Apenheul, then there need be few complaints.

But we can take nothing for granted. We should not casually assume that every modification effected in the name of behavioural enrichment is necessarily beneficial. Animals may ignore their new toys. They may play with the new devices for the first few days, and then forget them. The modifications may even do harm. It is important, then, to observe the responses; and indeed to measure them.

Does behavioural enrichment really work?

That devices for behavioural enrichment may do harm – and that the harm may be far from obvious – emerges from a salutary example in an American zoo. Chimpanzees were given a puzzle: a thin metal box with shelves inside, and with holes in the front to insert a finger. The top shelf harboured peanuts, and the chimps had to push the nuts down, shelf by shelf, by poking their fingers through. At the bottom, the nuts dropped out.

The first observation was that some individuals operated the puzzle far more than others. Those that used it most were the dominant animals. Their own general behaviour improved: they were less aggressive to their fellows, and showed less stereotypy. But the subordinate animals fared badly. They were frustrated because the dominant animals would not let them near the puzzle, and their stereotypies grew worse. Of course, one is tempted to suggest that a little applied anthropomorphism makes all this obvious. Do children like every toy that is given to them? Don't they grow bored? Don't they sometimes quarrel and sulk if the biggest child commandeers the best toys?

Of more direct benefit, though, is to measure the responses – employing the methods that have been worked out by field biologists throughout the century. For example, if there is some undesirable (or 'negative') behaviour that the keeper wants to reduce – for example some stereotypy – then he should measure its occurrence before and after introducing something to correct it. By the same token, of course, increase in desirable ('positive') behaviour can also be assessed, as Shepherdson measured the increase in gibbon brachiating and general activity after they heard the recorded songs. There are various statistical methods for ensuring that the comparisons are valid: for example, recording whether or not a particular behaviour occurs in particular time slots before and after the change is made. In general, it is important to make observations at different times of day (chimps, for example, may poke a termite nest all morning but leave it alone completely in the afternoon), and to repeat them a suitable number of times – where 'suitable number' is determined on statistical grounds, but generally means 'a lot'. If the statistical rules are followed, the results are sound.

Yet, as we keep stressing, the ultimate aim must be to return animals to the wild; not perhaps those of the present generation, perhaps not for many generations to come, but sooner or later. As we have already seen in chapter 5, animals that live simple lives – such as Arabian oryx – may take to the wild relatively easily. But as we have also seen, even oryx need to *learn* a great deal more than has hitherto been appreciated, before they can survive in the wild; and some animals – such as primates in forests – have more to learn than herbivores in deserts. It already seems clear that monkeys raised in a 'natural' environment, such as the Monkey Sanctuary's woollies, can return safely at least to quasi-wild environments. But conditions in captivity are not always as helpful as they are at Cornwall, and captive animals in general may need to be retrained for a wild existence. These are early days in the science of retraining. This is a new kind of problem. But there is already some progress.

TRAINING FOR THE WILD

The behaviourists finally convinced themselves that animals are machines, constructed to respond to external stimuli in pre-determined ways; and ethologists spoke of 'instincts' and 'drives', which again seemed designed to ensure that each animal 'automatically' behaved in ways that would promote its own survival, if placed in its habitual setting. In modern computer parlance, we would say that animal behaviour was perceived in very large part to be 'hard-wired'.

One of the two great revelations of the past few decades is that in practice, all but the simplest animals have to *learn* a high proportion of the skills they need for survival; and they do this as humans do, by being placed in the appropriate context at the appropriate age; by trial and error; by observing others; and indeed through direct instruction from their parents, siblings and fellows. Of course – as is also true of human beings – each kind of animal has only a limited repertoire: each kind does tend to do particular things in particular contexts, and in particular ways. But the repertoire is one of potential, rather than of in-built response and activity. We could perhaps draw a parallel with the human acquisition of language. Chomsky and Terrace argue that humans are born with an innate predilection for language: a desire to communicate (Terrace) and an innate ability to process sounds into sentences (Chomsky). Nevertheless, we have to learn how to speak a particular language, if this predilection is to be turned into a useful survival skill. Furthermore, if we want to be truly eloquent, then we need to be exposed to people who are truly articulate; and if we want a language to become 'native', as opposed merely to being proficient, then we should learn it from the cradle.

The same is true of the variety of abilities that animals have. In general, too, animals have to learn certain fundamental skills in order to survive: how, in general, to hunt or forage. But they also have to learn particular traditions that are appropriate to their own area. We have already seen, for example, how the Arabian oryx released into Oman steadily built up their knowledge of the local topography; and primates released into forest must learn where particular food trees grow, and at what times they are in fruit. Such knowledge is passed from individual to individual and generation to generation within each population: 'tradition' is a highly appropriate term. Bright animals such as chimpanzees may also hit upon brand-new skills every now and again which are also passed from generation to generation, producing traditions expanded by inventiveness. Thus 'tradition', in the brighter animals, truly becomes 'culture'. The second revelation of recent decades, as we have seen is that the lives of animals, and the repertoire of behaviours that enable them to survive, are far more complex than had been appreciated.

So – if animals are brought up in captivity, they have to learn or relearn the skills of the wild if they are to be returned safely. A few years ago it would have seemed strange even to contemplate that a captive animal *could* learn its 'natural' skills – and even stranger to suppose that they could learn much of what they need to know from human beings. Indeed it transpires, hardly surprisingly, that some species, some individuals within species, and some age groups, learn more easily than others. In general, though, it is beginning to seem that most captive-raised animals can learn enough at least to get by in a quasi-natural, protected environment; and although very few reintroduced animals have yet gone beyond the first generation in the wild, there is good reason to believe that their offspring, born in the wild, will be able to survive in even wilder conditions. Indeed, we may reasonably hope that as time and the generations pass, reintroduced groups will acquire much or most of the skill and behaviour of their original, wild ancestors – or at least will acquire enough of the ancient knowledge to be able to keep going.

Hence, training captive animals for the wild has become an accepted part of modern conservation practice, taking place before release or after, and preferably both. The general principle – as with all good teaching – is to begin simply and become more complicated; to work step by step from what the pupil knows, to what it does not.

Exactly what is required in a retraining curriculum was outlined by Dr Hilary Box of Reading University in her address to the Zoological Society of London's symposium on reintroduction in 1989.[10] She summarised the skills that animals need to learn under five headings: orientation; feeding; places to rest and sleep; relationships between species; and relationships within species. The structure of the following account is based upon her review, though with diversions.

The arts of moving and orientation

Animals that live in trees, such as most primates, have to learn how to move about in them. Each individual has to learn to trust its weight (and when to trust its weight) to branches that spring back, and vines that swing in strange trajectories; and if it has been brought up in a cage of rigid iron supports, it finds this far from simple. Orang utans released untrained from captivity are likely to refuse to climb trees at all. One of the obvious ways around this is to ensure that the enclosures of primates (in particular) do contain natural vegetation. Climbing skills cannot be learned too early!

Problem two is to learn orientation; for the canopy is complex, a three-dimensional puzzle offering a seeming infinity of routes; and the ability to move through a single tree, or from tree to tree, economically and without disaster, is a miracle of mental mapping. Animals that grow up in simple boxes, with only one practical route from one side to the other, may never acquire the necessary physical and mental skills. Mental mapping of the local topography is an additional problem; how to learn and remember which parts of the environment contain which kinds of things.

One obvious approach to such problems is to raise primates not in cages but on islands, as Jersey Zoo and many others do these days – or indeed to keep them virtually free, as at Apenheul. When Alison Hannah released chimpanzees into a reserve in Liberia, she actually showed the animals around the place! The ones that were introduced later learnt the way around from the ones already there. At Apenheul, too, tamarins and marmosets who were born in more restricted environments used only part of their range – but the next generation used a wider area. And so on.

Finding food

Animals newly released to the wild must also learn to find and catch food. A whole range of abilities are required: discrimination (What is worth eating?); 'optimum foraging strategy' (How much time is it worth spending to find each kind of food? How much risk is it worth taking?); to the simple ability to find or catch what is needed. Under each heading, the animals are born with potential abilities but each potential must be developed by learning. Cats are clearly born with a predilection to chase whatever moves, but not with the experience to dispatch particular prey efficiently. A female cheetah with cubs will catch a gazelle and then let it go – so her offspring can perfect the arts of bringing them down.

Omnivorous animals such as rats and most primates (including humans) face the problem of discrimination writ large. For rats and primates virtually any component of the environment could, in theory, be good and useful to eat; 'you just can't tell till you try'. However, most components of the environment (such as earth and wood) are not worth eating; some are frankly toxic; but a minority (which may still be a very large number!) are nutritious, though some growing leaves are nutritious when you first start to eat them, but become more toxic as they are consumed (because many plants respond to attack by rapid manufacture of toxin).

Omnivores have the most complicated diets, and in general, omnivorous mammals such as rats and primates (which are also intelligent) cobble together a repertoire of preferred foods from out of the plethora that surrounds them: thus human hunter-gatherers in tropical forests may regularly consume about 80 plants (plus a great variety of animals) from out of the many hundreds that grow round about. Some of the plants among this 80 (or however long the list is) are liable to be somewhat dubious; while other plants that are theoretically more nutritious may be left untouched. This is not perversity; it merely reflects the difficulty of the task. In practice, the repertoire can be added to only if some adventurous soul tries something new; but each experiment may be the adventurer's last. This is one aspect of the issue I discussed in chapter 1; that survival depends on a sometimes difficult compromise between conservatism and exploratoriness. Omnivores seem to be particularly alert to what others animals eat, however, and they copy successful experiments. So over time, each individual builds up its own repertoire, based partly upon its own but mainly upon the fruits of other individuals' adventurousness. Typically, we find that different populations of the same kind of animal (or different human tribes) living in very similar areas may none the less have a different repertoire of diet; because each, in the past, has experimented with different things. Typically, too, we find that animals (including humans) are most adventurous when they are young; but that once their range of tastes is acquired, they become highly conservative. Thus any of us would have developed a taste for seal blubber if we had been brought up among traditional Inuit, but if we were not, we would probably find the idea of it somewhat off-putting. We can see that this balance of adventurousness and conservatism is a highly efficient survival strategy. If we were not adventurous at all, then we would starve; but once a repertoire of food is established that is compatible with survival, then the safest course is to stick to it and to forsake all other.

Anyway, this balance of adventurousness (particularly in the young) and conservatism is precisely what is observed in wild omnivores or versatile vegetarians or insectivores, such as primates. We find, too, (as we might predict) that different populations of forest primate differ in the details of what they eat, because of their different traditions. But animals brought up in captivity have to learn what is good and bad to eat, in any particular home range. In addition, like hunters, they have to learn to cope with the difficulties raised by particular foods. No 'instinct' will tell a chimp that a termite mound harbours tasty tit-bits. It has to copy others; for it is extremely unlikely that any one individual would discover the skill for itself. No instinct can tell a chimp or a monkey that a particular tree by a particular stream produces fruit only at certain times of year.

Fortunately (as Hal Markowitz observed in a different context) animals in general like to fend for themselves. Alison Hannah's chimps in Liberia quickly began to eat wild fruit and leaves, even though there were back-up rations available. Other skills need special training: studies in Senegal and Gambia have shown that chimps can be taught (by humans) to probe termite mounds. Of course, animals introduced later can learn what they need to know from the experienced residents. But individual animals learn much more, and learn more quickly, from individuals they like: they pay more attention to them, and spend more time with them. It pays, then, to release animals in groups that can be seen to get on well with each other. Different generations and sexes, too, learn different things at different rates; and it has been found that golden lion tamarins in mixed sex and age-groups learn faster than male–female adult pairs.

A place to rest and a place to sleep

When we think of nests we think of birds: but mammals can be great nesters, too. Many a rodent builds a nest, of course, but so do many large animals. Chimps, gorillas, and bears build nests for sleeping in at night; while pigs build nests for raising their young. Despite the remarks above about the need for animals to learn a great deal, it is also true that they may retain a surprising amount instinctively, even without training. Thus domestic pigs build nests for their piglets if they are given the opportunity – even though many generations of their ancestors may have been raised in crates and styes. But animals in complicated environments such as forests do need to learn the good places to nest and to sleep – which includes the need to be out of the gaze, and preferably out of the reach, of predators. They also need to practise the skills of nest-building. The first nests of weaver-birds are, in general, a mess.

Relations with other species – especially predators

Island animals that have evolved over many generations in the complete absence of serious predators are often ridiculously tame, and fall foul of the first aggressor. The booby bird is a gannet, a marvellous flier with a beak like garden shears, but it acquired its name (from the Spanish for 'clown') because it happily landed on passing ships, and casually allowed itself to be thrust into the cooking pot.

Continental animals, that have evolved in the midst of predators, have developed a set of avoidance procedures which they lose to varying degrees in captivity, but which certainly need polishing up if they return to the wild. The golden lion tamarins that Devra Kleiman first released into reserves in Brazil in the early 1980s could not easily find their way through the trees (through lack of orientation skills) and tended to run along the ground instead. Wild individuals rarely do this; and in a land of jaguarundis (among other things) this is highly dangerous. The woolly monkeys at the Monkey Sanctuary retain their general wariness of predators (they duck when buzzards fly overhead, and avoid the sanctuary cat) but if they are reintroduced to the wild, they would need to learn the specific skills associated with the avoidance of big arboreal snakes, and of the harpy eagles that scour the canopy.

An over-confident attitude to humans is a hazard of captive-born animals in two ways. If they are too tame, they are in theory easily hunted; though primates at least seem highly susceptible to the norms of their social group, and we have already seen that a hand-raised woolly monkey released at Noah's Park in Brazil rapidly became one of the least tolerant animals, in its group, of human beings. Perhaps more trouble-some, though, is that captive-born predators may not realise that it is in everyone's interests for them to stay away from humans. Thus a cheetah released from De Wildt near Pretoria had to be brought in again after it showed a predilection for chasing motor-cyclists; and another turned up on a farmer's verandah, which is charming, except that cheetahs can devastate sheep and can be dangerous for children. In general, the feral dogs that run around many a city in southern Europe and the Third World are far more dangerous than wolves, which have learned that it does not pay to molest human beings, or even to look as if they were interested in molesting them. As an aside, we may note that badly-trained domestic dogs can be *far* more dangerous than wolves, first because they have no tradition of avoiding humans, and secondly because they are bred to be stupid and hyper-aggressive. Wild animals are aggressive only when they need to be. Yet we shoot wolves, and allow people who know nothing about handling dogs to take big and often neurotic creatures into public parks.

Can captive animals really learn the ways of the wild?

The general answer to that general question must be 'yes'. We have already seen it happen hundreds of times. Most instances so far involve domestic animals that have escaped, but as we saw in chapter 5, there is a growing catalogue of reintroductions for conservational purposes some of which already look convincing. To be sure, some present more problems than others; but problems are there to be overcome.

To date, though, perhaps the most striking vindication of captive breeding comes from Noah's Park in Brazil and – again – the Monkey Sanctuary. Marc van Roosmalen of Noah's Park is seeking to stock his reserve (a former capybara farm) with native, local animals.

These local animals include woolly monkeys. Endangered they may be, but rubber tappers kill the adults for meat, and sell the infants in the local market as pets. One baby woolly is worth a month's wages. The trade is illegal; but intervention seems pointless, once the babies reached the market, because if they were confiscated at that stage there would be nowhere to put them. Noah's Park, however, does provide a suitable sanctuary.

The trouble is that the baby woollies are too young to cope in the wild. By the end of 1990, however, the Monkey Sanctuary had three males of 4 years in age, Nick, Ricky, and Ivan, (we have already met them in chapter 5) who were beginning to get to the age when they should lead a group of their own; but the resident dominant male, Charlie, still had many years left in him, and the older Django was established as the number two. For genetic reasons, it seemed that Nick should stay in Cornwall, but the Sanctuary's head keeper, Rachel Hevesi, hatched a plot with Marc van Roosmalen to take Ricky and Ivan out to Noah's Park to act as 'tutors' to the group of babies that had been gathered together in late 1990.

At the time of writing, it is not clear whether the experiment will work. There is every reason to hope that it will, however. For instance, Rachel Hevesi has been able to foster a very young woolly captured from the market on to a still immature female that was already established in the Park. This female took the baby from Rachel, but would return it when called to be fed. Yet this female was not 'tame'; indeed she is one of the Park woollies who never normally comes down from the trees. The general picture is, indeed, that wild-living woollies can establish a relationship with humans that is distant enough to conserve their wildness, but also is close enough to be invoked *ad hoc* for the benefit of the monkeys.

In Cornwall, Ricky and Ivan have already shown that they cope well with infant monkeys, playing with them and allowing them to ride on their backs, in the typical woolly manner. They should be able to impress on the wild-born monkeys the general skills of living, while at the same time, the babies – who were born in the area – should retain good local knowledge of the available food. All the animals should retain enough tolerance of humans to allow themselves to be helped if necessary – but without being dependent on human help. If the experiment succeeds, then its significance will be profound; for this, surely, will be the first occasion on which captive-born animals have been used to tutor and chaperone wild-born animals to help them to survive in their own environment.

Twenty years ago, such an exercise as this would have been considered ridiculous. Now it seems a logical extension of what has already been achieved. That is progress. Indeed we have witnessed progress throughout this book; in ecological theory; in husbandry; in breeding strategy; in management of the wild; in understanding of animal psychology; in relations between human beings and animals. Where might such progress lead?

EIGHT

THE FUTURE

There have been times in the history of the world, and even of humanity, when nothing much happened. The fossil record shows that some snails remained unchanged through tens of millions of years, and prevailed throughout in much the same numbers; suggesting that their environment remained much the same as well. For a million years our own ancestor, *Homo erectus*, apparently altered very little, and made much the same tools at the end as at the beginning. However long they may prevail, however, such intervals are fragile. A single change can trigger another change, and then another, and another, and suddenly (relatively speaking) the entire environment is altered, and many or even most of the prevailing creatures go extinct. So far (as we saw in chapter 1) we know of five such mass extinctions in the past 600 million years.

Always though, there were a few creatures that survived the mass extinctions of the past. It is impossible to say, even in retrospect, why some particular animals survived, and others did not; there does not seem to be any identifiable quality of resilience. All we know is that some creatures did pull through, which is why we are here now to write about it.

Ours, beyond any question, is a period of transition. It is tempting to draw a parallel between what is happening now, and the mass extinctions of the past. But as we saw in chapter 2, the parallel is not quite apt, for the mass extinctions of the distant past seem in part at least to have crept up on living things, perhaps over millions of years. We have apparently been exterminating large animals at a prodigious rate over thousands of years; and now we threaten to eliminate at least half the remaining species on Earth within a few decades simply by obliterating tropical forests. In scale, then, the mass extinction that we are now perpetrating is comparable with those of the past; but in speed, the present turmoil far outstrips anything that has happened before. Because it is so much faster, there is no time for animals to adapt; the option of survival by evolving into something else, which some creatures were able to do in the past, is not available. Because we have interrupted the landscape, too, reducing plains and forests to sequestered pockets, we have removed the option of migration, which in the past enabled many species to escape, for example, the worst of the Ice Ages.

Always in the past, too, mass extinction was brought about by changes in climate, perhaps exacerbated by changes in topography as continents coalesced or divided, or by impact of meteors. Never before have entire suites of species disappeared simply because one particular creature pushed them aside. On the grand scale, without the agency of agriculture, that is an ecological impossibility. But the present mass extinction is brought about directly or indirectly by the single species, *Homo sapiens*. To say that we are in control of the extinction would be to flatter us. But we are the cause, and we can influence the course of events, by choosing to do some things, and not others. The outcome of the present extinction is to some extent dependent upon human will.

The mass extinctions of the past have all been followed, once things had time to get into their stride, by a period of *adaptive radiation*. The animals that chanced to survive often evolved subsequently with astonishing rapidity and variety, to fill the ecological niches that had been vacated. The dark ages were followed by renaissance. Thus it was that the mammals first appeared on Earth 200 million years ago – at about the same time as the dinosaurs. For the next 130 million years or so they lived in the shadow of the dinosaurs, as small, hole-in-the-ground or arboreal creatures similar in general form to modern tree-shrews. But within a few tens of millions of years of the dinosaurs' extinction the mammals radiated miraculously – into whales and mega-carnivores and recognisable elephants, mice and seals. (The bat evolved before the dinosaurs disappeared, and when they had gone diversified to become among the most various of mammals, second only to the rodents.)

If the present obliteration of wildlife is at all analogous with the mass extinctions of the past, then we might hope for a comparable renaissance some time in the future. It will begin, of course, as all have done in the past, with whatever survives the period of decline. The trouble is, from the human point of view, that although the time required for adaptive radiation to take place is a twinkling of an eye in geological terms – a million years or less can bring remarkable change – it is very long by human standards. The present extinction is *ours*; we brought it about by our efforts, and most of it has occurred and will occur in a few decades. Perhaps it is proper for us to engineer the renaissance as well, for if all we leave behind us is rats and cockroaches, it will take too long to build again, and might never again achieve the erstwhile diversity.

Thus, I see present conservation endeavours in two contexts. The first is to limit the extent of the mass extinction. The second is to set in train the renaissance; to ensure that when the world returns again to a constructive phase, analogous with the renascent period of adaptive radiation, there is plenty to build upon.

The two exercises are complementary, and they must proceed side by side, for there is only one world that we are concerned with, and only one strand of time to work within. Habitat protection (I think it could be said) can be seen as an exercise in damage limitation. Captive breeding, and other attempts to sustain diversity above what can now be achieved in the wild is, by analogy, the foundation of the renaissance.

However, even if we accept captive breeding as a legitimate pursuit – which of course is the contention of this book; does this justify the *zoo*?

DO WE REALLY NEED ZOOS?

If we define 'zoo' narrowly, as a collection of animals in cages with bars, then the answer to that grand question is 'no'. But zoos, as we have agreed throughout this book, do not have to be like that; should not be like that; and – if we are talking about good, modern zoos – are not like that. If we define 'zoo' more broadly – as a place where animals live in a protected state, and are made accessible to human observation – then the zoo becomes self-justifying. If zoos did not exist, then any sensible conservation policy would lead inevitably to their creation.

Indeed, if we survey modern methods of conserving animals, we see an entire spectrum of approaches. At one extreme is the intensive breeding centre. This does not have to be (and should not be) a row of cages; but if animals are kept in small spaces (and any artificial space is liable to be small compared with the wild) and yet are breeding, then this by definition is 'an intensive breeding centre'. At the other end of the spectrum is the wilderness itself. In between, is every kind of compromise: fenced reserves ('sanctuaries') for just one species; tightly managed reserves with a select list of species and natural vegetation; reserves with natural vegetation in contact with 'wilderness', but with protection from predators; the national park, which resembles wilderness, but must none the less be managed to maintain its diversity and prevent local extinctions; and wilderness itself, in which the ecology is allowed to unfold entirely according to its own rules.

Wilderness remains the 'ideal', the dream, the representation of the pristine world. Wilderness must be allowed to prevail wherever possible (albeit visited and studied by human beings); and conservationists should endeavour to push as far as they can from the intensive end of the spectrum, towards the wilderness. Zoos in general should grow into national parks, and some national parks at least might eventually become so large that they can (almost) be left alone.

The spectrum exists: and each component of the spectrum can be justified in the present world, provided that each is good of its type. Various adages make the point: 'horses for courses'; 'more than one way to crack a nut'. In short, there are advantages and disadvantages in each kind of sanctuary. The urban zoo, though its space is small, makes very good use of available cash; and cash is at a premium in conservation. Urban zoos, tiny though they are compared with continents, nowadays between them sustain populations that compare in size and in a few cases outstrip those of the wild. Single species sanctuaries (or zoos) can be good because they concentrate effort, and because conditions can be adjusted to a particular species needs. Of course there is a case for sanctuaries for black rhinos, and for breeding cheetahs at De Wildt. But mixtures of species can be helpful too: grazing animals in general make better use of herbage if complementary species graze together; woods have canopies for monkeys, but also leave floor-space for deer and pigs; the scientists who study one species of cat can learn from looking at others – and through comparisons with totally different creatures; the vets who master the care of one exotic species use their time best if they extend their expertise to others.

The fact that there is a spectrum of approaches – as opposed merely to a variety – is helpful too. Reintroduction is the proper end point of conservation breeding. As we have seen in chapters 5 and 7, this is enacted best by introducing animals to environments that become wilder by degrees. As things are, they can be moved from the intensive breeding centre to the 'natural' enclosure, from the enclosure to the reserve, and so on. Such a phased reintroduction cannot take place until all the intermediate stages exist.

Why, though, keep animals in Chester or Cincinnati that properly live in India or Brazil? If they have to be protected, why not keep them in reserves in their own country? Of course there is a powerful case for conserving animals at home. Black rhinos in reserves in Kenya and South Africa; Sumatran rhinos and Bali starlings in Indonesia; European bisons in Germany; these and hundreds more are examples of legitimate *in situ* conservation. But as we discussed in chapter 2, the same problems that beset an animal in the wild can also threaten it in captivity, if it stays at home. Elephants in reserves in Vietnam, or scimitar-horned oryx in Chad, would not have escaped the ravages of war. The population of Puerto Rican parrots in a reserve in Puerto Rico was halved by Hurricane Hugo. Populations *in situ* are highly desirable – necessary, indeed – but it also makes sense to keep other populations elsewhere.

Finally, as we observed in chapter 1, the reason for conserving animals is not simply that they are good for us. It is at least a worthwhile metaphor to regard our fellow creatures as another part of Creation, of which we are the most privileged members. More broadly, we should conserve animals because we should want to conserve them, and should want to conserve them because it is right to do so.

Nevertheless, it would be perverse and gratuitously damaging to them and to us not to take every opportunity to benefit from their existence. To put the matter crudely, conservation in poor countries (and conservation in general is most urgent in poor countries) must as far as possible pay its way. In rich countries, too, people must be persuaded to part with cash; for nothing succeeds in the present world without it.

Most of the ways in which we gain profit from animals are to some extent unpleasant for them. Some involve killing the animal, and some (as in circuses) involve its degradation. But if we are content simply to look at animals, and to be in their presence – which really ought to be reward enough – then it should be possible to gain from them without upsetting them. In practice, observation by more than a few specialists is not so simple to arrange, because animals do not go out of their way to be looked at. Yet with modern technology, a lot of ingenuity, and a great deal of money, it is possible to reveal animals to large numbers of people without disturbing them. The techniques range from closed-circuit television in the eagle's nest at Cincinnati, to carefully conducted tours in national parks (as opposed to jamborees in land-rovers). Jeremy Cherfas discusses a great many possibilities in *Zoo 2000*. The point, though, is that such revelation is legitimate; and indeed is necessary, partly because it raises revenue, and partly because it does increase public awareness. Unless people care about animals, there is no hope for them; and they will care more for them if they are able to see them.

We will return to the matter of looking at animals. My only point here is to show that the zoo – defined broadly – invents itself. If you protect animals (as is proper) and show them off to people (as is necessary to arouse public sympathy and to raise revenue) then you have, by definition, a zoo. It is clear, too, that if conservation is to succeed over the next few centuries, in which the human population will reach its peak, then we (and animals) will need more and more protected places; while humans, who perhaps will have more leisure, and rising expectations, will want more and more access to them. In short, the zoo (defined broadly) is necessary. It will be part of civilised living for the foreseeable future.

Change, though, is inevitable. In general, and obviously, the spectrum of reserve from breeding centre to national park will become blurred; it will become more and more difficult to decide what is a zoo, and what is a reserve with public access. Definitions do not matter a great deal. All that matters is that whatever exists is right for its circumstances; and that the trend, in general, should be to make places wilder, rather than tamer; that the ideal role of human beings is not to organise the play, but simply to ensure that the play can take place.

Neither, in the future, will it be so easy to distinguish between the population that are wild, and those in captivity.

WILD POPULATIONS AND ZOO POPULATIONS

As we discussed at length in chapter 4, numbers are a crucial issue in conservation. Populations that are too small go extinct. We saw, too, though, that viability is not simply a matter of numbers. If all the animals in a population are genetically similar, or if each subpopulation is inbreeding, then that population is in danger; unless, like the cheetah, it contains very few deleterious alleles, and so can tolerate homozygosity. Yet we have seen that the remaining wild areas are often too small to accommodate viable populations; and that zoos, too, will have increasing difficulty in maintaining genetic variation within their inevitably limited numbers.

Part of the answer to both these problems is to maintain a flow of animals, or their genes, between the captive populations and the wild ones. At present, there is very little flow – although it is already difficult to draw a clear distinction between the wild and the captive animals within the golden lion tamarin population of Brazil. Increasingly, though, through the techniques discussed in chapter 6 – the collection of semen from the wild, or (eventually) of eggs from dead females; and the transfer of embryos – we will see a flow of germ plasm from wild to captivity, and from captivity to the wild. In short, wild and captive populations increasingly will become genetically continuous; just as different zoo populations that are members of the same species survival plan, are already genetically continuous.

Increasingly, too, it will become feasible (it is already technically possible) to keep track of particular individuals in the wild, to take blood or other tissue samples from them without upsetting them, and to analyse their DNA. Even in the short term, it will thus be possible to perceive that a particular wild individual is genetically very dissimilar from a particular captive population, and hence (assuming it is not too distant, and is a different subspecies) that it will provide a useful antidote to inbreeding. Eventually, it could become possible to show that particular stretches of DNA are, in fact, particular genes with a particular function; and hence to identify wild individuals who contain precisely the genes that the captive population has lost.

Eventually, indeed, it will be possible (and considered desirable) to enact the same processes in reverse; perhaps to observe that a particular wild population is losing genetic diversity too rapidly, and to add to it from the captive pool, either by inseminating females in the wild, or (where this is socially possible, and not too dangerous) by reintroducing particular individuals into the wild population.

In short, the genetic management that is now practised within the best-run captive breeding plans will be extended to embrace the wild populations as well. Yet this should be done only when it is necessary. Perhaps in an ideal world both the wild and the captive populations would be so large that neither would need to borrow genes from the other; though then, perhaps, it might be worth exchanging genes simply to ensure that the two populations do not 'drift' apart, and evolve into separate subspecies. But future generations of biologists can cross those bridges when they come to them; and be grateful, that they have such a dilemma to solve.

The demarcation between zoos and more extensive reserves will blur, then; and so, too, will the distinctions between the wild and the captive gene pools. As time goes on, too, we will increasingly see the loss of demarcation between the professional conservators of animals, and the amateur.

PROFESSIONALS, AMATEURS, AND COMMERCIAL BREEDERS

Throughout this book I have taken it for granted that species survival plans or EEPs should be carried out in zoos or reserves that are linked or at least affiliated to some 'official' body, and in particular to the IUCN; and that they should in general be run by professional zoologists, keepers and vets. But I have also pointed out first that all conservation endeavours are chronically short of cash, and second that it is extremely helpful and indeed vital in all conservation endeavours to engage the interest of people at large.

Clearly there is a conflict here. Conservation needs professionals, of course. But professionals add to the cost; and by being professional, they are inclined (sometimes intentionally and sometimes inadvertently) to cut themselves off from the laity. Excellent zoological societies with exemplary conservation records have sometimes appeared to the world at large not as public institutions, but as private clubs. But there are two loose and heterogeneous groups of people who are not in general 'official' – by which I mean in this context that they are not officially sanctioned by the IUCN or some comparable body; but who could, in theory, help enormously to alleviate both of the principal problems that the professionals face. The first of these groups are the 'amateurs', or 'hobbyists'. The second are the commercial breeders, who sell animals both to hobbyists and to professional zoos. Amateurs are sometimes bungling incompetents and many have no serious interest in conservation at all. Some commercial breeders are both incompetent and cynical, and would prefer animals to be rare so as to increase the cash value of the ones that remain. But the ranks of the amateur and the commercial breeder also include some of the finest of all keepers of animals – people who truly have 'green fingers'. Many are conscientious, dedicated, and care deeply about the survival of species. Between them, too, amateurs and commercial breeders spend and handle truly prodigious sums of money: amateur aquarists spend £100 million a year on their hobby in Britain alone, and about US$15 billion worldwide. Furthermore, too, amateurs by definition *are* 'the people'; by engaging them, the gap between the pro and the public at large is bridged at a stroke. For all these reasons, many 'pros' now feel that the 'resource' of the hobbyist and the breeder should be tapped far more diligently.

The principle, of course, is well established. Many a ranch in the USA harbours antelope from Africa and Asia; herds that could and in some cases do augment the 'official' captive conservation herds. Auctions are held in South Africa for ranchers to bid for surplus 'game' from reserves. This may sound distasteful. But ranchers that I have seen (on television) say that they do not feel that they 'own' the animals; merely that they have bought the privilege of looking after them, and conserving them for the future. Such a sentiment seems to me to be unimpeachable; and the sheer size of some American and African ranches compared with that of many reserves suggest that their contribution could be very great indeed. Père David's deer was of course saved by the Duke of Bedford on his estate at Woburn, and Arabian oryx raised in palace gardens in the Middle East contributed to the original 'world herd', and continue to broaden the genetic pool.

Of course, dukes and American ranchers are not exactly the hoipolloi. Their involvement does not quite provide the democratisation that seems desirable. But lesser spirits (financially speaking) could also do a great deal. We spoke in chapter 4 of the need for bachelor herds: ideal for people who have space for a few animals in a hectare or so, but do not want the trouble of breeding. Malayan tapirs are an endangered species: a dozen or so males dotted around the gardens of English country hotels or the rustic headquarters of commercial companies, could make an enormous difference. Private owners could also, for example, maintain animals that are perfectly fit but are not wanted for serious breeding; perhaps because their genes are already well represented, or perhaps because they are hybrids. Private owners then would simply provide sanctuary; but this is vital, because many a breeding programme is being held up because zoos are reluctant to liberate space by euthanising present incumbents. The only caveat would be that bachelors or animals given sanctuary might thrive at the expense of native fauna and flora; but this would not apply if they were kept, say, as an alternative to sheep (of which Europe as a whole has far too many).

Yet you do not need a big house in the country to contribute to conservation. Aquarists and aviarists (and even keepers of reptiles and amphibia) often achieve remarkable results in private apartments and garden sheds. Late in 1990 Dr Gordon Reid of the Horniman Museum in South London, who maintains some of the only remaining breeding groups of some of the Lake Victoria haplochromines (chapters 2 and 5) was negotiating with the British Cichlid Society to help him in his endeavours. People who can raise other kinds of cichlid (members of the family Cichlidae) can also keep haplochromines. Invertebrates – perhaps particularly – could benefit from 'amateur' assistance. Paul Pearce Kelly of London Zoo is wondering at this minute whether and to what extent to elicit outside help in rearing (by hand) the multitudinous progeny of his red-kneed tarantulas.

Parrots as a group are appallingly endangered, as we discussed in chapter 5. Some commercial dealers in parrots are enemies of the order, and indeed of life; they pluck them from the wild, and welcome rarity to increase cash value. But some breeders are among the most skilful and knowledgeable bird handlers in the world, and some are dedicated conservationists. Harry Sissen in Yorkshire for example has wonderful success with macaws and in 1989 alone he and his wife bred Palm Cockatoos, lesser Vasa Parrots, Hawk-heads, and Blue-throated and Golden Conures. Very few zoos compare.

There are theoretical dangers in the involvement of amateurs and breeders. The tradition among many aquarists, for example, is to 'improve' their protégés – like the breeders of 'koi' carp, with their perpetual quest for exotic colours. They stop at nothing; some deliberately create hybrids so as to produce new forms, and fish in general hybridise easily (because, as we have seen, there is often very little genetic difference between related species, even though the ecological differences may be profound). As we saw in chapter 4, selective breeding and the creation of hybrids are anathema to the conservation breeder. In general, then, hobbyists or commercial breeders who wanted to be involved in species survival plans would have to accept the discipline of the species coordinator, just as zoos do. Many will surely find, though, that the satisfaction of belonging to a serious conservation endeavour exceeds the pleasure of breeding what in many cases are simply grotesques.

Some professionals fear, too, that amateurs may simply lose interest, and abandon the task half-way through. We all of us do, of course; we all grow tired of hobbies. On the other hand, no one would entrust an entire species to just one keeper, amateur or a professional. Neither can we be sure that professionals are 100 per cent reliable. Scientists run out of grants, zoos close down, reserves get flattened by hurricanes. Nothing is safe. We just have to spread the risks as best we can.

Finally, it has been suggested that amateurs might require more organising than their endeavours would be worth. This, though, seems merely to be a theoretical point. Good amateurs really are good. A few hours' input from a professional now and again could elicit hundreds of hours[2] (and hundreds of thousands of pounds') worth of help. Indeed the time could come – and why not? – when amateurs might also act as keepers of studbooks or even as species coordinators. For retired people this could be a wonderful occupation: absorbing, and granting instant membership of a worldwide cause. You do not have to keep animals to be a good organiser.

In short it seems reasonable to regard the possible involvement of amateurs and breeders in the same light as the cooption of the Harasis as wardens of Arabian oryx. Everybody benefits; and because everybody benefits, the animals are more secure. These, then, are the practical changes that could and should be made in the future, as the conservation endeavour evolves, and as zoos settle into their new roles. But we also need changes in attitude: changes in human attitudes, of course; but also interestingly – changes in the attitude of animals.

HABITUATION

If animals are to survive in viable numbers in a world crowded with human beings, then we must adjust our attitude to them. Equally, though, but inescapably, *they* must adjust their attitude to *us*. What they must acquire is that special state of mind that lies between what we mean by 'wild' and 'tame'. It is the state exhibited by robins in surburban gardens, who watch from the edge of the wheelbarrow yet would never be enticed into the house; the state that biologists call habituation.

The state of habituation is eminently explicable in behaviourist terms. An animal responds to a stimulus: but if the response is not reinforced (by a reward) or punished, then, after a time, the animal ceases to respond at all. All animals can habituate. Snails learn quite quickly not to withdraw their 'horns' in response to a puff of wind – if the puff is repeated often enough and is not followed by any more serious ill effects. Soldiers in war learn to ignore the general clatter of gunfire. But habituation nonetheless is a subtle state; for soldiers never lose their ability to 'hear' and to respond to the special sound of a shell that is targeted upon them. The brain filters the incoming stimuli, which is a useful thing to do; but it does not switch off.

The evolutionary value of habituation is equally easy to explain. All animals that are not in a state of sensory deprivation are constantly assailed with stimuli to which they could theoretically respond. An animal that responded to everything would be a nervous wreck. A bird that flew away from every susurrus and shadow would never settle long enough to feed. The animal's nervous system must be equipped with filters, therefore; and repeated stimuli that denote nothing dangerous must be ignored. A balance must be struck, however. A tame pigeon in the country, or a soldier who is habituated to the point of carelessness, are both liable to be shot.

The state of habituation is flexible. An animal that is habituated in one context will not necessarily be habituated in another. Wood pigeons in cities will feed from your hand. Carrion crows hop among the deck-chairs in St James's Park, in the heart of London. If you confront either of those two birds in the country, you would not easily get within a quarter of a mile. In the country, people shoot them; or at least, they have done so often enough in the past to establish a tradition of avoidance. Of course, you could argue that the habituated pigeons of the city and the elusive pigeons of the wild are different subpopulations; that they have not learned to habituate or to be nervous, but have simply been subjected to natural selection. The idea here would be that habituated individuals in the country would simply be shot, while nervous individuals in the city would never settle to feed. I know of no information on this matter, with respect to crows and pigeons; but evidence from other species shows that individual birds clearly can adjust their 'attitude' to suit the circumstance. Thus greylag geese amble around the public parks of Scandinavia just as Canada geese do in London. But when the greylags migrate, as they do, to southern Europe, *the same individuals* become difficult to approach. In southern Europe, people shoot them.

Pigeons, crows, and geese, however, are easy examples. Some people don't like pigeons and some don't like crows, but there is no particular reason to be afraid of them. Mega-carnivores offer more of a challenge: a cheetah on the stoop is off-putting if you have small children (or sheep), and it must be a little disconcerting to watch one casually gaining ground as you work through the gears on your motor-bike. As we have observed, too, there is no more dangerous animal on Earth than a large and aggressive domestic dog; precisely because it has no fear of human beings.

But big wild carnivores are intelligent: a lot more intelligent than geese which, as we have seen, are perfectly able to adjust their behaviour from place to place. Wolves still live in highly populated countries: Italy, Iran, Germany, the United States. Occasionally people have been frightened by wolves but *no one* in living memory has ever been killed by one except, very very occasionally, in remote places by animals with rabies. The fault there, though, lies with the virus rather than the animal; and it is theoretically possible to control rabies in wild animals by leaving bait that has been laced with oral vaccine. As Hartmund Jungius of IUCN once put the matter, wolves have 'manners' – an aspect of their traditions; they know they should not threaten human beings. At present, though, wolves are more wary than seems ideal; it would be good if they were habituated to the point where they were happy to be seen, but not to the point at which humans and wolves are liable to come into direct contact, for that might lead to misunderstanding. We need not underestimate them, however. They are capable of learning.

Biologists working in the field commonly find that after a time, even the most elusive creatures become habituated to them. The gorillas of Rwanda are still wild animals; but they do not scatter when people come to see them, if the people keep their distance. The woolly monkeys in Noah's Park vary in their behaviour: some will come down from the trees to meet people they know (such as Marc van Roosmalen) and some will not. They are habituated enough to be visible, and they are relaxed in the presence of humans; they are not forced into human company in a state of terror. But they are not tame. They would not stand around like booby birds and allow themselves to be stroked or fondled, or put in a cage, or snaffled by an eagle.

Of course we cannot be stupid about this. The bears that used to hang around the camps in North American national parks were extemely dangerous. If they had not been habituated (and positively encouraged) they would not have been near the camps. Such meetings, though, are not what I have in mind. It would probably be equally stupid to encourage wolves into villages with tins of dogfood, because once they are in a human settlement and out of their home territory, they are liable to feel trapped if taken by surprise, and to respond in panic. The point is not to invite confrontation, but simply to reduce the aversion that now seizes most wild creatures when they get wind of human beings. Other species have a healthy respect for each other, when respect is appropriate. But they are not cut off from each other, in the way that most of them are from us.

If animals are to become habituated, however, then we must learn also to adjust the way that we respond to them. The contact that most people have with animals these days is with domestic pets, and in particular with dogs; and this can give a false impression of the kind of relationship that is in general realistic. Dogs are very special, partly because they are social animals; partly because their particular society is very hierarchical, with subordinate animals being positively fawning (to the point where the subordinates do not even breed); and partly because – after thousands of years of breeding and association – they seem willing to think of themselves as honorary humans, and of humans as honorary, dominant dogs.

Most animals do not have this particular biology and history. Most animals, most of the time, do not perceive advances by other animals of different species as an attempt to establish friendship; and they do not perceive the need for such friendship. In so far as they establish social relations with their own species, these are conducted through particular sets of signals, which can be peculiar to each particular species. Humans are more curious than most other species, and more acquisitive; and, like other primates, we are highly tactile. We tend to stare at the things that interest us, and to want to touch them. But most animals, most of the time, would see such approaches as a threat. Even with animals that are genuinely tame, and like to be handled, it is easy to get the wires crossed and threaten when we mean to be kind. Woolly monkeys, for example, can become very tame; and the tame ones like to be handled, for they too spend much of the time in physical contact with their fellows. But if one woolly monkey places a hand on the head of another woolly monkey, this means, 'I am superior to you, and you must give way.' For humans, the pat on the head and the stroking of hair are signals of affection and approval. This worries Erich Mager at Apenheul; that a visitor and one of the tamer woollies will one day misunderstand each other. When monkeys feel threatened, and they cannot escape, they bite. In general, if we want monkeys or wolves or any other animal to be polite to us, then we must learn to be polite to them: respect their dignity; learn their codes; give them 'space'.

Taken all in all, indeed, we must learn a relationship with animals that is not one of ownership or of patronisation, but is one of respect and appreciation: closer to what true connoisseurs, as opposed merely to collectors, feel in the presence of great art. The appreciation of animals has a broader dimension than this, for animals are sentient creatures and our fellow beings, as inanimate objects are not. Yet the concept of connoisseurship is worth considering.

CONNOISSEURS OF ANIMALS

A theatre ticket in a big Western city is likely to cost three to 20 times as much as a ticket to the zoo. Many zoos enjoy some tax concessions because they are charities, but most depend for their income on people coming through the gates, plus the subscriptions of members, plus whatever legacies and sponsorship come their way. Very few zoos have more money than they could put to good immediate use, and many struggle from year to year. We seem simply to have grown used to the idea that the zoo just offers a cheap day out, and that its role is merely to tire the chidren.

Things do not have to be like that; and if zoos truly contribute to conservation, as they can, and if animals are truly valuable, as they are, then things certainly should not be like that. For me (as for all the keepers I have ever spoken to) zoos are a necessary part of any itinerary. Whenever I travel I find out if there is a zoo worth visiting en route. In the nature of my job I am able to wangle the odd trip: Hanover was the first I consciously manoeuvred with a zoo in mind, in 1976; Washington via Cincinnati has been the most recent. I have also arranged holidays around zoos: in 1990, to Holland and Germany, taking in Amsterdam, Arnhem, Apeldoorn, Stuttgart, and Cologne. In Britain, there are a dozen or so zoos to which I make fairly regular diversions. The point of all this is not to bore you with memoirs or even with holiday slides but simply to suggest that zoos are good to add to the itinerary, along with all the other reasons for visiting new places.

Some of the zoos I have visited have been tatty, pointless, and miserable. Bad ones still exist. But I seek out the good ones and worldwide, or even in some individual countries, they form an impressive cadre. I have often been privileged, and taken on conducted tours by directors, or scientists, or keepers. Partly for this reason, but partly, too, as a paying visitor, I have enjoyed scores of excellent hours. Details tend to stay in the mind: the family of red pandas emerging from its den in a November dawn at Cincinnati: dawn again at the Monkey Sanctuary, Cornwall, and threatened in the drizzling rain by Charlie, the alpha male; San Francisco, where the baby musk ox trotted in rapid convoy, like manic little squads of soldiers, as wart-hogs do; Glasgow, in the black-bears' den: Chester, close encounters with the elephants – but also with the capybara, reaching from the lake to browse the weeping willows; the paddle fish at London, with their gill arches spread like trawl nets, to filter plankton; summer evening at Marwell, among the antelope on a Hampshire hill after the visitors had gone; the cheetahs and vultures at De Wildt; the free-range Madagascan orb-web spiders in Washington, and the amazing anemones, shrimps, and chitons. 'What a piece of work is a man', said Hamlet. What a piece of work is *any* creature, if you take the time to look; and zoos can be good places to look.

Such times as this are available to everyone. It is only a question of going often, to different places; and of returning to the same haunts at different times, in different moods. If zoos become a hobby, part of what you do with your life, then it becomes reasonable to spend time and money on them. If you belong to the local zoo society (and most good zoos do have societies of 'friends' and 'fellows') then you can go as often as you want; and it is often best just to go for an hour or so at a time, sometimes to look at particular animals, and sometimes just to walk around.

Zoos, in their turn, could do far more for people who genuinely like to look at animals, and take an interest in them. Many zoos now have volunteer guides, some of whom are extremely well informed. In some countries – Britain, for example – it is not usually easy to find someone who can truly take time to show you what is really happening in the zoo, and has the necessary insight, unless you have the excuse of writing a book. Many zoos, too, have beautifully designed labels these days; but very few indeed tell you what you really want to know: why these particular animals are in this particular enclosure; how they are kept; which individuals dominate which; whether the particular population is part of a coordinated captive breeding plan; what is the status of the species in captivity and in the wild. The information conveyed by most zoo labels is of a far more general kind; often the kind of thing you could find in any children's encyclopaedia. In short, very few zoos really do enough to encourage that minority of people (I would say an important minority) who genuinely would take the trouble to become connoisseurs.

Consider what might be achieved. People in England and a few comparably eccentric places watch a game called cricket. Major games are scheduled for five days. Even comparatively run-of-the-mill en-counters last three or four days. Nothing much happens for hours at a stretch, yet people who watch it achieve a state of mind that is quite addictive, a quiet contemplation punctuated by moments of ecstasy. My brother and I, as boys, used to watch all three days of a three-day match. Animals too can be watched in just the same way. Professionals do watch them in such a vein when in the field; and amateurs too may already spend hours in hides watching water fowl, or (for the richer) on balconies that overlook selected (and carefully contrived) water-holes in Africa. I wonder, though, why we might not watch bears (for example) in cities, in places contrived for them: parks with look-out posts, and perhaps with a grand-stand by some focus of activity. In such places it would be possible to breed animals in significant numbers, and retain their behaviour; and yet indulge the interests of people who are no more eccentric than the spectators of cricket matches, to watch for hours at a time when the fancy took them.

There are plenty of possible approaches. What matters in the short term is that animals should survive in sufficient variety, and with enough of their behaviour intact, to take them back to the wild at some future, more relaxed and more enlightened time. Zoos are helping that to happen, even as things are. If zoos and all other serious endeavours in conservation had ten times their present resources then huge numbers of people could gain far more enjoyment and enlightenment from contact with animals in the short term; and we could begin to be optimistic about the long term. In the future, too, if animals became habituated through lack of threat, then we could expect to enjoy an association with them that was far closer and more frequent than at present; a kind of *Arcadia*. To invest ten times the amount that we spend at present would not be too much for such a prize. For rich countries the total amount would still be trivial. When we consider what we do spend money on, and what could be achieved if we spent it sensibly, it seems like nothing at all.

At their best, even at their present best, zoos defined broadly can be acceptable places for animals to live for generation after generation. If an animal lives as a member of a social group that is proper for its species, feeds from natural vegetation, finds its own mates and rears its own offspring, it does not seem to matter over much if it is also protected from predators, or is amenable to human observation. Even the best zoos, though, should be and should be seen to be only a part of a spectrum, that extends from the intensive centre to the wilderness. They are not the sole end point of conservational endeavour, but for the forseeable future they must be seen as part of the endeavour.

In a few centuries' time, if all goes well, the importance of zoos will diminish. For the present, their significance must grow. For the next few centuries, they must be perceived as a necessary part of civilisation. When they rise to this challenge, they become exciting. Any lesser perception, or lesser endeavour, is a trivialisation.

NOTES

AUTHOR'S NOTE

Though I was educated as a zoologist, my working methods are those of a science writer. Much of what I know and remember is taken from meetings, taped interviews (not least during days at BBC Radio), and conversations – many with directors, curators, scientists, vets, and keepers, while gazing into paddocks. This *modus operandi* has advantages; but I regret that I cannot supply worthwhile written references for every quote, because in many cases they do not exist. This is particularly true for chapter 8, where all the quotes are from people I have met or talked to on travels.

The following references are to books that I have found particularly valuable, or otherwise illuminating; meetings, which ought to be available in proceedings in the fullness of time, if not by the time of publication (details are obtainable from the societies cited); and a selection of key papers.

INTRODUCTION

1. Frankel, O.H. and Soulé, Michael E., *Conservation and Evolution,* Cambridge University Press, 1981.

ONE

1. Myers, Norman, Ed., *The Gaia Atlas of Planet Management,* Pan Books, 1985.
2. Jones, David, *Lifewatch,* London Zoo & Whipsnade, Spring 1990, p. 10.
3. *New Scientist,* 3 March 1990, p. 22.
4. Harris, Marvin, *Good to Eat,* Allen & Unwin, 1986.
5. Tudge, Colin, Ed., *The Environment of Life,* Oxford University Press, New York, 1988.

TWO

1. May, Robert, 'How Many Species Are There on Earth?', *Science,* vol 241, 1988, pp 1441–9.
2. Quoted in May, *op cit.*
3. See note 1 above.

4. Quoted from Darwin's letters by Fisher, R.C., 'An Inordinate Fondness for Beetles', *Biological Journal of the Linnean Society,* vol 35, 1988, pp 131–319.
5. See note 1 above.
6. Myers, Norman, Ed., *The Gaia Atlas of Planet Management,* Pan Books, 1985.
7. Diamond, J.M., 'The Present, Past and Future of Human-caused Extinctions', *Evolution and Extinction,* The Royal Society, London, 1989.
8. Collar, N.J. and Andrew, P., *Birds to Watch. The ICPB World Checklist of Threatened Birds,* International Council for Bird Preservation, Cambridge, 1988.
9. See note 7 above.
10. See note 8 above.
11. See note 7 above.
12. See note 7 above.
13. McKenna, Virginia, Travers, Will and Wray, Jonathan, *Beyond the Bars,* Thorsons Publishing Group, 1987.
14. See note 13 above.
15. Soulé, Michael, et al, *Zoo Biology,* Alan R. Liss, Inc., New York, vol 5, 1986, pp 101–114.
16. See note 7 above.

THREE

1. *International Zoo News* published six times a year by Zoo-Centrum, London.
2. Hofmann, R.R. and Matern, B., 'Changes in Gastrointestinal Morphology Related to Nutrition in Giraffes, *Giraffa camelopardalis*: A Comparison of Wild and Zoo Specimens', *International Zoo Yearbook 27*, Zoological Society of London, 1988.
3. See for example Goodall, Jane, *In the Shadow of Man,* rev ed, Weidenfeld & Nicolson, 1988.
4. Brambell, Michael and Matthews, Sue, *The Zoological Society of London 1826– 1976 and Beyond*, Academic Press, London and New York, 1976, pp 147– 65.
5. Scott, Patricia P., 'The Special Feature of Nutrition of Cats, in Crawford, M.A., Ed., *Comparative Nutrition of Wild Animals*, Zoological Society of London/Academic Press, 1968, pp 21– 36.
6. Crawford, Michael and Crawford, Sheilagh, *What We Eat Today*, Neville Spearman, 1972.
7. Toone, William D. and Risser, Arthur C. Jr, 'Captive Management of the California Condor', *International Zoo Yearbook 27*, Zoological Society of London, 1987, pp 50–58.
8. Quoted in Tudge, Colin, 'A Wild Time at the Zoo', *New Scientist*, 5 January 1991, pp 26–30.
9. Cherfas, Jeremy, *Zoo 2000*, British Broadcasting Corporation, 1984.
10. Gould, Nicholas, *International Zoo News* No 222, Zoo-Centrum, July/August 1990, pp 2–3.
11. Williamson, Henry, *Tarka the Otter*, Putnam & Co. Ltd., 1927.

FOUR

1. Darwin, Charles, *The Origin of Species* (18), Penguin Books, 1983, p. 76.
2. Huxley, Julian S., *Evolution: The Modern Synthesis* (19), new edn, Chatto & Windus, 1963.
3. Foose, Thomas J. and Ballou, Jonathan D., 'Population Management: Theory

and Practice', *International Zoo Yearbook 27*, Zoological Society of London, 1988, pp 26–41.
4. Conway, William, 'The Practical Difficulties and Financial Implications of Endangered Species Breeding Programmes', *International Zoo Yearbook 24/25*, Zoological Society of London, 1986, pp 210–219.
5. Mayr, Ernst, *Towards a New Philosophy of Biology*, Havard University Press, Cambridge, Mass., 1988.
6. de Boer, Leobert, *EEP Co-ordinators' Manual, May 1989 Version*, National Foundation for Research in Zoological Gardens, Amsterdam, 1989.
7. Ferguson, A., 'Conservation of Genetic Diversity in Brown Trout and Other Salmonids'. Paper presented at 'The Biology and Conservation of Rare Fish', The Fisheries Society of the British Isles Symposium, Lancaster University, UK, 16–20 July 1990.
8. Ryder, Oliver A., et al, 'Individual DNA Fingerprints from Galapagos Tortoises', *International Zoo Yearbook 28*, Zoological Society of London, 1989, pp 84–7.
9. See Meffe, G.K., 'Genetic Approaches to Conservation of Rare Fishes: Examples from North American Desert Species, *Journal of Fish Biology*, Academic Press, 1990, pp 105–112.
10. O'Brien, Stephen, 'A Molecular Solution to the Riddle of the Giant Panda's Phylogeny', *Nature*, vol 317, 1985 pp 140–4.

FIVE

1. Scheel, David and Ross, Dina, 'How Not to Save a Species', *New Scientist*, 16 October 1986, pp 39–42.
2. Kawata, Ken, 'Japan's Survival Programme Gets Off the Ground', *International Zoo News*, January/February 1991, pp 6–8.
3. St Paul, Corinthians 13:13.
4. Horizon, BBC 2, 3 March 1991.
5. Brett, R.A., 'The Status of Sanctuary Populations of the Black Rhinoceros in Kenya'. Report for the Gallmann Memorial Foundation, Nairobi, Kenya,

and the Institute of Zoology, Zoological Society of London, June 1990.

6. Grimwood, Ian, 'Operation Oryx: The Start of It All', in Dixon, Alexandra and Jones, David, Eds., *The Conservation and Biology of Desert Antelopes*, Christopher Helm, 1988, pp 1–8.

7. Kolter, Lydia and Zimmermann, Waltraut, *EEP Co-ordinators' Manual, May 1989 Version*, National Foundation for Research in Zoological Gardens, Amsterdam, 1989.

8. Glatson, Angela, 'The Red Panda Studbook', *International Zoo News*, vol 223, September 1990, pp 5–8.

9. Powell, Roy, *Paignton Zoological and Botanical Gardens Newsletter*, Summer 1990.

10. Maitland, Peter, Comment at 'The Biology and Conservation of Rare Fish', The Fisheries Society of the British Isles Symposium, Lancaster University, UK, 16–20 July 1990.

11. Stanley Price, Mark, 'A Review of Mammal Reintroductions'. Paper at 'Beyond Captive Breeding', Zoological Society of London Symposium, 24–25 November 1989.

12. Laycock, G., *The Alien Animals*, Ballantine Books, New York, 1970.

13. Lever, Christopher, *Naturalized Birds of the World*, Longman Scientific and Technical, 1987.

14. See note 11 above.

15. See note 6 above, pp 14–17.

16. Williamson, Doug, 'Gulf Gazelle Reintroduction', *Lifewatch*, Winter 1990, pp 12–13.

17. Kleiman, Devra, in Benirschke, Kurt, Ed., *Primates: The Road to Self-sustaining Population*, Springer–Verlag, 1986, pp 959–79.

18. Moore, Donald and Smith, R., 'The red wolf as a model for carnivore reintroduction'. Paper at 'Beyond Captive Breeding', Zoological Society of London Symposium, 24–25 November 1989.

19. See note 1 above.

20. Toone, William D. and Risser, Arthur C., 'Captive Management of the California Gymnogyps californianus', *International Zoo Yearbook 27*, Zoological Society of London, 1988, pp 50–57.

21. Cherfas, Jeremy, 'Return of the Native', *New Scientist*, 11 March 1989, pp 50–53. 'The Value of Reintroduction to Bird Conservation', Report of symposium organised by The Wildfowl Trust and the International Council for Bird Preservation, 29 November – 1 December 198?.

SIX

1. For a summary of research and ideas of the Institute of Zoology, as alluded to throughout this chapter, see *Science for Conservation*, Zoological Society of London, 1991.

2. Dresser, B.L., 'Embryo Transfer in Exotic Species'. Paper at 'Biotechnology and the Conservation of Genetic Diversity', Zoological Society of London Symposium, 4–5 September 1990.

3. See for example Fehlly, C.B. Willadsen, S.M. and Tucker, E.M., 'Interspecific Chimaerism between Sheep and Goat', *Nature*, vol 307, 1984, p. 634; Meinecke–Tillmann, S. and Meinecke, B., 'Experimental Chimaeras of Sheep and Goat', *Nature*, vol 307, 1984, p. 637.

SEVEN

1. Medawar, Peter, *The Art of the Soluble*, Methuen, 1967.

2. 'Animal Intelligence'. Meeting of The Royal Society, 6 June 1984.

3. See note 2 above.

4. Markowitz, Hal, *Behavioral Enrichment in the Zoo*, Van Nostrand Reinhold Co., New York, 1982, p. 8.

5. Cherfas, Jeremy, *Zoo 2000*, British Broadcasting Corporation, 1984.

6. See Tudge, Colin, 'A Wild Time at the Zoo', *New Scientist*, 1991, January 5, pp 26–30.

7. Hediger, H., *Man and Animal in the Zoo*, Routledge & Kegan Paul, London, 1969.

8. Shepherdson, David, *Ratel*, vol 16, 1989, p. 100.

9. Markowitz, Hal, *Behavioral Enrichment in the Zoo*, Van Nostrand Reinhold Co., New York, 1982, p. 16.

10. Box, Hilary, 'Training Animals for Life after Release'. Paper at 'Beyond Captive Breeding', Zoological Society of London Symposium, 24–25 November, 1989.

INDEX